KT-442-026

EDEXCEL LEVEL 3

MATHS

IN

CONTEXT

PROJECT BOOK

CORE MATHS

PEARSON

Published by Pearson Education Limited, 80 Strand, London WC2R 0RL.

www.pearsonschoolsandfecolleges.co.uk

Copies of official specifications for all Pearson qualifications may be found on the website: https://qualifications.pearson.com

Text © Pearson Education Limited 2016
Typeset by Tech-Set Ltd, Gateshead
Original illustrations © Pearson Education Limited 2015
Index by Indexing Specialists (UK) Ltd

The rights of Nick Asker, Jack Barraclough, Ian Bettison, Huw Kyffin, Su Nicholson and Robert Ward-Penny to be identified as authors of this work have been asserted by them in accordance with the Copyright, Designs and Patents Act 1988.

First published 2016

19 18 17 16
10 9 8 7 6 5 4 3 2 1

British Library Cataloguing in Publication Data
A catalogue record for this book is available from the British Library

ISBN 978 1 292 14928 8

Printed in Slovakia by Neografia

Acknowledgements
The authors and publisher would like to thank the following individuals and organisations for permission to reproduce photographs:

(Key: b-bottom; c-centre; l-left; r-right; t-top)

123RF.com: kessudap 107; **Alamy Images**: FirePhoto 122, Paul Kingsley 110; **Digital Vision**: 96; **Fotolia.com**: Andriy Solovyov 105, ASampedro 66, Batman57 28, behindlens 119, furtseff 68, Itan1409 133, ktsdesign 74, nolonely 90, Rawpixel 2, romanolebedev 27, Romolo Tavani 104, Tomasz Zajda 72, カシス 78; **Getty Images**: artpartner-images 130, Lester Lefkowitz 86, Sinan Saglam / EyeEm 48, Yiu Yu Hoi 5; **Pearson Education Ltd**: Jules Selmes 93; **Science Photo Library Ltd**: Andrzej Wojcicki 116; **Shutterstock.com**: Annette Shaff 1, bikeriderlondon 60, Carlos Amarillo 52, David Pereiras 11, dibrova 81, Dmitry Bruskov 21, Eric Limon 19, Jose AS Reyes 10, leungchopan 36, Lynne Carpenter 82, Marafona 115, Marta Paniti 46, Mathias Rosenthal 8, Maxisport 39, Moreno Soppelsa 127, Nickolya 57, Paket 16, PavelSvoboda 92, Peter Clark 99, RUSOTURISTO 65, Singkham 51, spaxiax 54, Suzanne Tucker 33, testing 24, Wallenrock 75, Yellowj 126, Zhukov Oleg 45

Cover images: Front: **Corbis**: Blend / Roberto Westbrook; **Shutterstock.com**: Africa Studio; **Back**: Shutterstock.com: Aila Images

All other images © Pearson Education

We are grateful to the following for permission to reproduce copyright material:

Figures
Graph on p.12 data from The World Bank: Life expectancy at birth, total (years): Data source: Derived from male and female life expectancy at birth from sources such as: (1) United Nations Population Division. World Population Prospects, (2) United Nations Statistical Division. Population and Vital Statistics Report (various years), (3) Census reports and other statistical publications from national statistical offices, (4) Eurostat: Demographic Statistics, (5) Secretariat of the Pacific Community: Statistics and Demography Programme, and (6) U.S. Census Bureau: International Database http://web.worldbank.org/WBSITE/EXTERNAL/0,,contentMDK:22547097~pagePK:50016803~piPK:50016805~theSitePK:13,00.html.

Tables
Tables on page 13, pages 13-14 data from The World Bank: Life expectancy at birth, total (years): Data source: Derived from male and female life expectancy at birth from sources such as: (1) United Nations

Population Division. World Population Prospects, (2) United Nations Statistical Division. Population and Vital Statistics Report (various years), (3) Census reports and other statistical publications from national statistical offices, (4) Eurostat: Demographic Statistics, (5) Secretariat of the Pacific Community: Statistics and Demography Programme, and (6) U.S. Census Bureau: International Database http://web.worldbank.org/WBSITE/EXTERNAL/0,,contentMDK:22547097~pagePK:50016803~piPK:50016805~theSitePK:13,00.html; Table on page 19 adapted from Marriage Summary Statistics (Provisional) 2012, 10 September 2014, Table 2b, © Crown Copyright 2014. Source: Office for National Statistics licensed under the Open Government Licence v.3.0; Table on page 20 adapted from Marriage Summary Statistics (Provisional) 2012, 10 September 2014, Table 1, © Crown Copyright 2014. Source: Adapted from data from the Office for National Statistics licensed under the Open Government Licence v.3.0; Table on page 24 adapted from Overview of the UK Population, 5 November 2015, Table 1, © Crown copyright 2015, Source: Adapted from data from the Office for National Statistics licensed under the Open Government Licence v.3.0; Table on page 52 adapted from PAYE tax and Class 1 NICs, PAYE tax and Class 1 NICs, Tax thresholds, rates and codes, Contains public sector information licensed under the Open Government Licence v.3.0; Table on page 61 from Land Registry, http://landregistry.data.gov.uk/app/hpi/, Contains public sector information licensed under the Open Government Licence v3.0; Table on page 93 adapted from Typical Stopping Distances The Highway Code, Department of Transport, 2 October 2015, Rule 126, Source: Adapted from data from Office for National Statistics licensed under the Open Government Licence v3.0; Tables on page 99, page 100, page 101 adapted from Travel Trends (various editions), http://www.ons.gov.uk/ons/rel/ott/travel-trends/index.html., Source: Data from Office for National Statistics licensed under the Open Government Licence v3.0; Table on pages 123-24 adapted from Fire Facts, Fires in Greater London 1966-2014, 30 April (London Fire Brigade Information Management Team 2015) Tables 1.1 and 1.2, Contains public sector information licensed under the Open Government Licence v3.0; Table on page 129 from World Class OEE, www.oee.com/world-class-oee.html, originally published on www.oee.com and used with permission from Vorne Industries, Inc.

Text
Extract on page 83 adapted from FCA confirms price cap rules for payday lenders, https://www.fca.org.uk/news/fca-confirms-price-cap-rules-for-payday-lenders, © Financial Conduct Authority.

Microsoft Excel® is a registered trademark of Microsoft Corporation in the United States and/or other countries. Used with permission of Microsoft.

Websites
Pearson Education Limited is not responsible for the content of any external internet sites. It is essential for tutors to preview each website before using it in class so as to ensure that the URL is still accurate, relevant and appropriate. We suggest that tutors bookmark useful websites and consider enabling students to access them through the school/college intranet.

Contents

Welcome to the Edexcel Level 3 Maths in Context Project Book.

This book is packed full of projects you can work on individually or in groups to practise using maths in a real-life and realistic way.

Clear design helps prepare you for the **comprehension** and **application** elements of the qualification.

Unit openers show what you will cover in each unit, as well as the prior knowledge needed from GCSE.

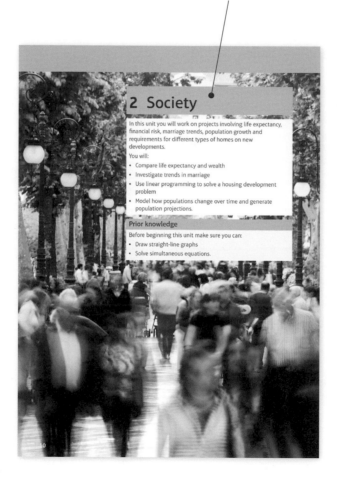

The 2014 Core Maths Technical Guidance from the Department for Education places a strong focus on preparing you for contexts you are likely to encounter in vocational and academic study and in future employment and life. The Edexcel Level 3 Certificate in Mathematics in Context builds on this approach to learning. Pearson has developed a suite of Progression Services which support teachers, learners and parents in understanding what progress looks like and provides support to track if you are on course to achieve your potential.

New for September 2016: the Pearson Progression Scale will include the Edexcel Mathematics in Context content descriptors. The Scale can be used to track your progress and to identify barriers to learning and opportunities to improve.

Visit www.pearsonschoolsandfecolleges.co.uk/mathsincontext for more information about the Maths in Context course, including how to access answers and lesson plans.

Visit www.edexcel.com/mathematicsincontext to access the full range of support from the awarding body.

Context-led principles so you learn and apply the maths you need when you need it.

Key points and **Examples** where you need them, for new Level 3 and Higher GCSE maths content.

2.4 Social housing

Data source A

A property developer is putting forward a proposal to build the Weston housing estate. The local council insists that this new housing development is designed in a way that combines privately owned homes with rented social housing. This 'mixed tenure' approach is intended to break up areas of relative poverty and bring different social groups together. The developer wants to make as much profit as possible while still following the council's requirements.

The developer knows from market research that in this area most people want to buy three-bedroom houses with small gardens. The local council has asked the developer to also include smaller two-bedroom flats for the rented social housing.

The plot of land can be broken down into 210 small blocks of building space. Each flat takes up two blocks of space and each house uses five.

The council is also concerned about traffic problems. It has therefore restricted the total number of new dwellings to a maximum of 72.

The council also insists that the number of flats is either equal to or greater than the number of houses.

The developer expects to make a profit of £20 000 on each flat and £60 000 on each house.

Technical literacy

A property developer arranges the construction of new properties or renovates existing buildings.

21

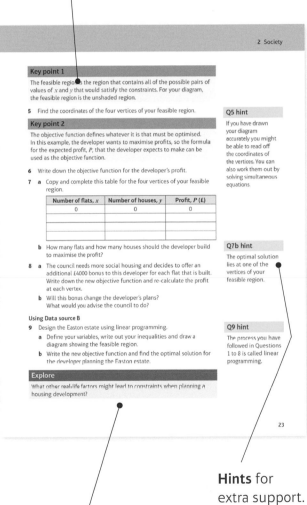

Key point 1

The feasible region is the region that contains all of the possible pairs of values of x and y that would satisfy the constraints. For your diagram, the feasible region is the unshaded region.

5 Find the coordinates of the four vertices of your feasible region.

Key point 2

The objective function defines whatever it is that must be optimised. In this example, the developer wants to maximise profits, so the formula for the expected profit, P, that the developer expects to make can be used as the objective function.

6 Write down the objective function for the developer's profit.

7 a Copy and complete this table for the four vertices of your feasible region.

Number of flats, x	Number of houses, y	Profit, P (£)
0	0	0

b How many flats and how many houses should the developer build to maximise the profit?

8 a The council needs more social housing and decides to offer an additional £4000 bonus to this developer for each flat that is built. Write down the new objective function and re-calculate the profit at each vertex.

b Will this bonus change the developer's plans? What would you advise the council to do?

Using Data source B

9 Design the Easton estate using linear programming.
a Define your variables, write out your inequalities and draw a diagram showing the feasible region.
b Write the new objective function and find the optimal solution for the developer planning the Easton estate.

Explore

What other real-life factors might lead to constraints when planning a housing development?

Q5 hint

If you have drawn your diagram accurately you might be able to read off the coordinates of the vertices. You can also work them out by solving simultaneous equations.

Q7b hint

The optimal solution lies at one of the vertices of your feasible region.

Q9 hint

The process you have followed in Questions 1 to 8 is called linear programming.

23

Hints for extra support.

Technical literacy hint boxes to explain real-life terms.

The **Explore** questions prompt you to take your learning as far as you can.

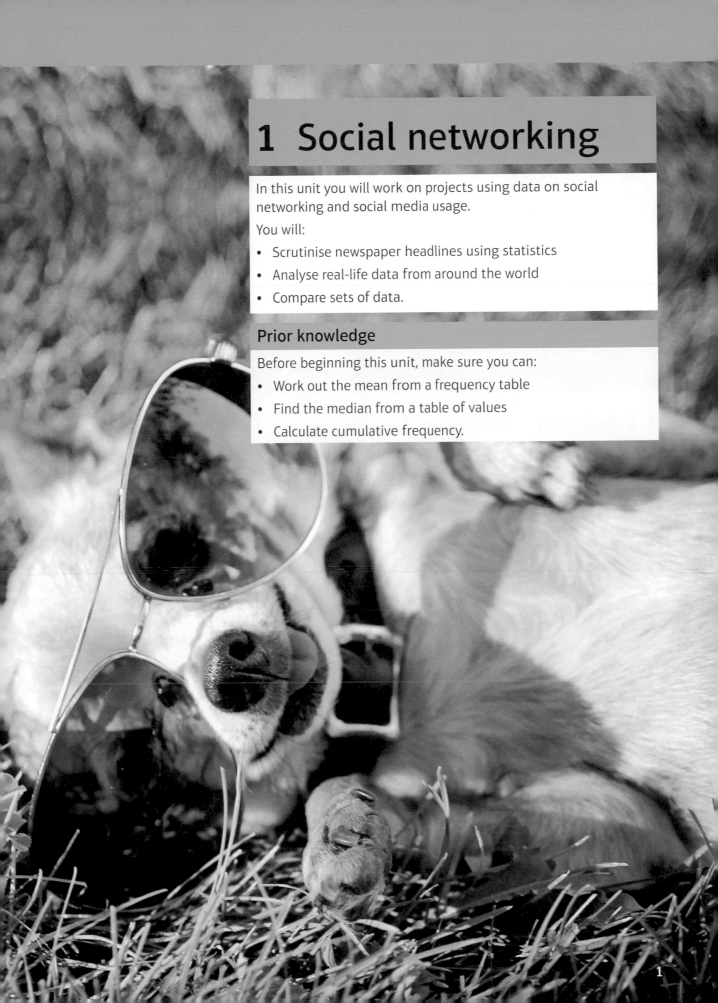

1 Social networking

In this unit you will work on projects using data on social networking and social media usage.

You will:

- Scrutinise newspaper headlines using statistics
- Analyse real-life data from around the world
- Compare sets of data.

Prior knowledge

Before beginning this unit, make sure you can:

- Work out the mean from a frequency table
- Find the median from a table of values
- Calculate cumulative frequency.

1 Social networking

1.1 Social network usage

Data source

A teacher at a sixth-form college wants to investigate how students use social media.

The college has 1400 students. The teacher asks a sample of 50 of them how many different social networking accounts they each have and how long they spend social networking during their lunch hour.

This list shows the number of social networking accounts that each student has.

6, 2, 5, 1, 8, 6, 4, 7, 8, 4, 7, 3, 3, 10, 9, 1, 2, 9, 8, 2, 8, 2, 2, 6, 9,
4, 5, 8, 8, 1, 9, 1, 5, 7, 5, 3, 9, 3, 6, 3, 8, 4, 5, 4, 3, 9, 5, 3, 6, 7

Table 1 shows information about the times that the students in the sample spend social networking during their lunch hour.

Table 1 Time spent social networking

Time, T (minutes)	Midpoint	Frequency
$0 < T \leqslant 4$	2	5
$4 < T \leqslant 8$	6	6
$8 < T \leqslant 12$	10	4
$12 < T \leqslant 16$	14	11
$16 < T \leqslant 20$	18	7
$20 < T \leqslant 24$	22	9
$24 < T \leqslant 28$	26	4
$28 < T \leqslant 40$	34	4

Table 2 shows values for the estimated mean and median and the upper and lower quartiles. The values were calculated using Microsoft Excel®.

Table 2

Estimated mean	16.08 minutes
Estimated median	14 minutes
Lower quartile	10
Upper quartile	22

Look at the data source

1   How could the teacher make sure that they choose a fair sample?
2   Find the mean number of social networking accounts using the data for the sample of students.

Key point 1

This formula can be used to find the mean of a set of data:

$$\bar{x} = \frac{\Sigma fx}{\Sigma f}$$

where

x is the variable

$\bar{x}$ is the mean of the frequency distribution

f is the frequency with which the value x occurs

Σ is the sum of the values. The capital sigma (Σ) means 'add all the values together'.

Example

Number of posts per day, x	Frequency, f	fx
1	5	$5 \times 1 = 5$
2	10	$10 \times 2 = 20$
3	2	$2 \times 3 = 6$

$$\bar{x} = \frac{\Sigma fx}{\Sigma f} = \frac{5 + 20 + 6}{5 + 10 + 2} = \frac{31}{17} = 1.82 \text{ (to 2 d.p.)}$$

3   Draw a cumulative frequency graph using the data in Table 1.

Q3 hint

Create a cumulative frequency table first.

4 Use your graph to estimate the median and the lower and upper quartiles.

Key point 2

For a set of n data values on a cumulative frequency graph:

- the estimate for the median is the $\frac{n}{2}$th value
- the estimate for the lower quartile (LQ) is the $\frac{n}{4}$th value
- the estimate for the upper quartile (UQ) is the $\frac{3n}{4}$th value.

5 The values in Table 2 were calculated using Microsoft Excel®. Compare these with the values you found in Question 4.

6 Here are two newspaper headlines.

College students have 10 social networking accounts each

College students spend over half their lunch break social networking

Use your findings to explain whether the teacher would support or contradict each headline.

Explore

How do these results compare to your own social network usage?

1.2 Social networking in different countries

Data source

Table 1 shows the number of individuals in the world using the internet in 2013 and in 2014.

Table 1 World internet users

Year	Number of world internet users (millions)
2013	2705
2014	2937

Source: Data from International Telecommunications Union

The gross domestic product (GDP) of a country measures the value of goods and services it produces.

The GDP per capita of a country is the GDP divided by the number of people in that country.

Table 2 shows every country that was part of the European Union in 2012, the country's GDP per capita and the percentage of individuals using the internet for social networking.

Here, social networking is defined as posting messages, blogs, discussions or other media to chat sites, newsgroups or similar sites.

Technical literacy

Per capita means 'for each head' – in this case, per individual.

Table 2 GDP and social networking

Country	GDP per capita ($ to the nearest thousand)	Percentage of individuals using the internet for social networking
Austria	48 000	37
Belgium	45 000	49
Bulgaria	7 000	30
Cyprus	29 000	37
Czech Republic	20 000	25
Denmark	58 000	48
Estonia	17 000	43
Finland	47 000	49
France	41 000	33
Germany	44 000	34
Greece	22 000	32
Hungary	13 000	43
Ireland	48 000	46
Italy	35 000	29
Latvia	14 000	37
Lithuania	14 000	47
Luxembourg	106 000	50
Malta	21 000	44
Netherlands	49 000	65
Poland	13 000	42
Portugal	21 000	45
Romania	8 000	22
Slovakia	17 000	47
Slovenia	22 000	45
Spain	29 000	42
Sweden	57 000	54
United Kingdom	41 000	57

Source: Based on information from the Office for National Statistics, the European Commission, Eurostat and the World Bank

Look at the data source

Using Table 1

1 Calculate the percentage increase in the number of individuals in the world using the internet between 2013 and 2014.

2 Assume that this percentage increase continues for the next few years and use it to predict the number of world internet users in 2020.

Q2 hint

You could use a table.

3 The world population in 2020 is estimated to be 7 643 402 123.

Use this information to comment on your predicted number of world internet users in 2020.

Using Table 2

4 Plot a scatter diagram to show the relationship between GDP per capita and the percentage of individuals using the internet for social networking.

  a Identify any outliers.

    Which country (or countries) do they represent?

  b What kind of correlation does the scatter diagram show?

5 A country outside the European Union has GDP per capita of $55 000.

  a Use your diagram from Question 4 to estimate the percentage of this country's population who use the internet for social networking.

  b Comment on the limitations of your estimate.

Q5a hint

Draw a line of best fit. Ignore any outliers.

6 Two politicians have opposing views on whether the use of social networking is related to GDP per capita.

Politician A considers that social networking boosts productivity and so increases the country's GDP.

Politician B takes the view that social networking is a distraction that lowers productivity.

Each politician chooses a sample of the countries to support their point of view.

Sample 1	Sample 2
Lithuania	Austria
Slovakia	Germany
Portugal	France
Spain	Italy
Germany	Czech Republic

  a Draw a scatter diagram for each sample.

  b Which sample did each politician use? Explain your answer.

1.3 Six degrees of separation

Data source

There is a theory that everyone in the world is connected to everyone else through no more than six people.

Henna is a sociology lecturer and she devises an experiment to find out how this theory holds up for her two new classes.

Before they meet, each student is given one of their new classmates' names.

They then use a Facebook app to find out how many connections there are between themselves and their new classmate. A connection of '1' means they know that person. A connection of '2' means one of their friends knows that person, and so on.

Table 1 shows the number of connections for each student.

Table 1 Number of connections to new classmate

Class 1		Class 2	
Student	**Number of connections, x**	**Student**	**Number of connections, x**
Katrina	1	Jean	6
Abdul	1	Paula	4
Erica	3	Francis	2
Heidi	1	Gustavo	4
Maxine	2	Laurence	3
Reatha	4	Naomi	2
Garrett	3	Richard	2
Diane	8	Alpita	5
Jacinda	1	Christopher	3
Jessica	2	Sarah	4
Nickolas	3	Tom	2
Christine	10	Tyrone	3
Renee	2	Claire	4
Yasuko	3	Laurence	2
Justin	6	Penny	3
Marguerite	1	Malena	2

Look at the data source

1   Work out the mean number of connections per person for
    a Class 1   **b** Class 2.
2   Look at the two sets of data. Which data set is more consistent? Explain how you made your decision.

Key point

The variance measures how spread out (or *varied*) a set of data is.
You can work out the variance using this formula:

$$\text{variance} = \frac{\Sigma(x - \bar{x})^2}{n}$$

where $\bar{x}$ is the mean of the set of data and n is the number of items of data.
Squaring the difference between each data value and the mean removes any negative values.
The standard deviation of a set of data is a measure of how far the values vary from the mean.
The standard deviation is the square root of the variance:

$$\text{standard deviation} = \sqrt{\frac{\Sigma(x - \bar{x})^2}{n}}$$

3   **a** Complete this table to work out the standard deviation for Class 1.

Class 1		
Student	**Number of connections, x**	$(x - \bar{x})^2$
Katrina	1	
Abdul	1	

Q3 hint

Look at the formula:

$$\sqrt{\frac{\Sigma(x - \bar{x})^2}{n}}$$

Find the sum of the squares of all the differences.

Then divide by n.

Then take the square root.

   b Work out the standard deviation for Class 2.
4   Classes 1 and 2 have the same mean but their standard deviations are different. What does this tell you about the data?
5   A new student from overseas joins Class 2.
    A student chosen at random in Class 2 finds that there are 20 connections between them and the new student.
    How will this new data value affect the mean and the standard deviation for Class 2?
6   Henna was expecting the mean number of connections to be less than 6. Why?
7   What assumption has Henna made when developing this experiment?

Explore

The six degrees of separation theory was developed in the 1920s but has never been proven. Research how different people have tested it.
Some people think that if the theory was developed now, you would need fewer than six connections. Why?

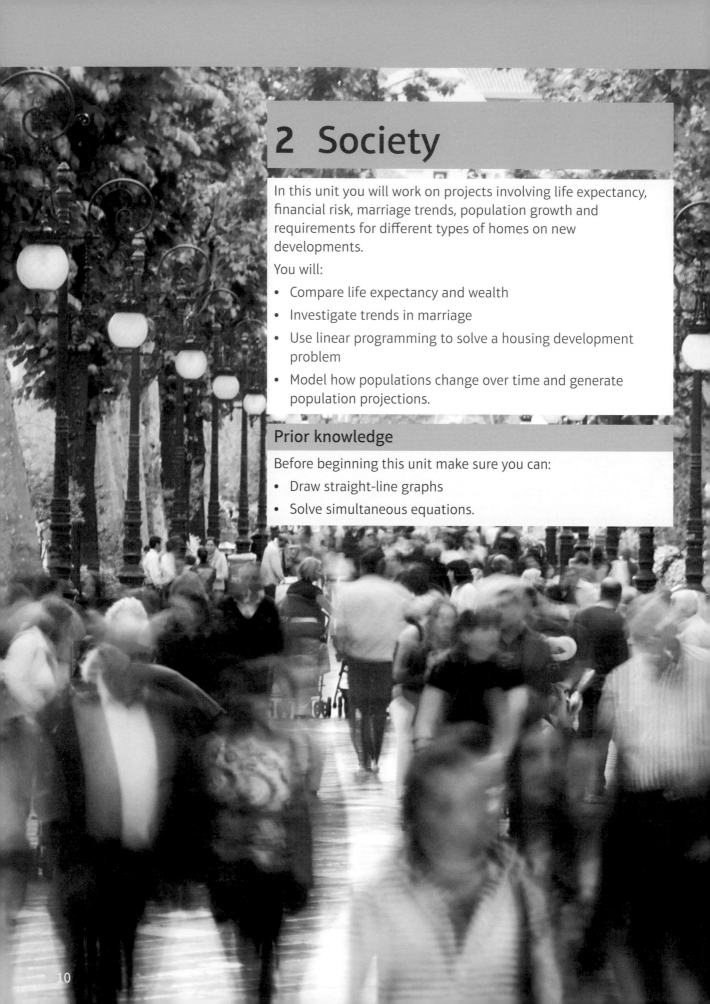

2 Society

In this unit you will work on projects involving life expectancy, financial risk, marriage trends, population growth and requirements for different types of homes on new developments.

You will:

- Compare life expectancy and wealth
- Investigate trends in marriage
- Use linear programming to solve a housing development problem
- Model how populations change over time and generate population projections.

Prior knowledge

Before beginning this unit make sure you can:

- Draw straight-line graphs
- Solve simultaneous equations.

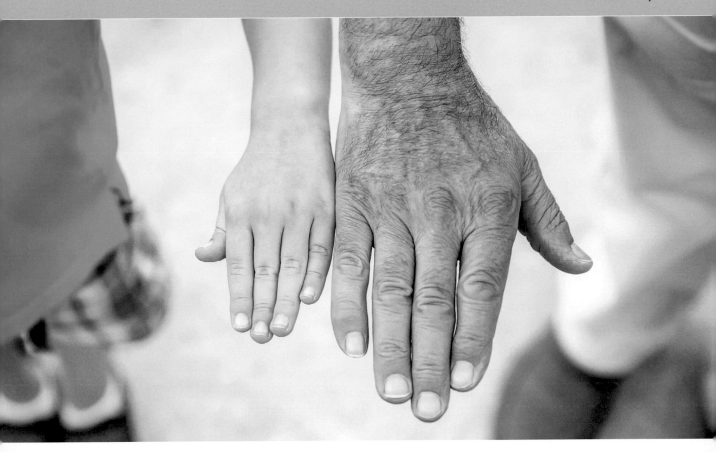

2.1 Life expectancy and wealth

Data source

Life expectancy is a statistical measure of how much longer, on average, someone of a given age can expect to live, assuming that mortality rates do not change. Graphs 1 and 2 show how life expectancy in England and Wales has varied since 1841.

Graph 1 Male and female life expectancy at birth in England and Wales, 1841–2009

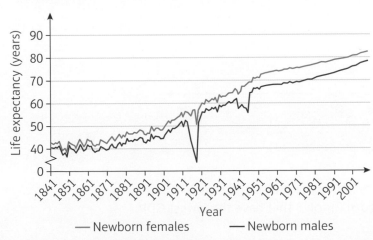

Source: Based on data from Human Mortality Database

Graph 2 Male and female life expectancy at age 65 in England and Wales, 1841–2009

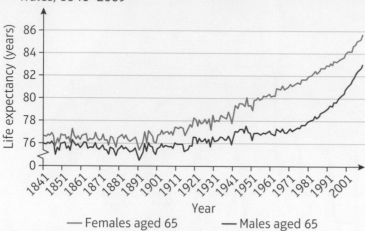

Source: Based on data from Human Mortality Database

Life expectancy at birth varies between different countries. It is as low as 45 in Sierra Leone and as high as 83 in Japan. Research shows a connection between the wealth of a country and the life expectancy of its people. Gross Domestic Product (GDP) per capita is the most common measure of how wealthy a country is. Graph 3 shows the connection between life expectancy at birth and GDP per capita in a sample of countries from around the world in 2012.

Graph 3 Life expectancy at birth against GDP per capita in 2012

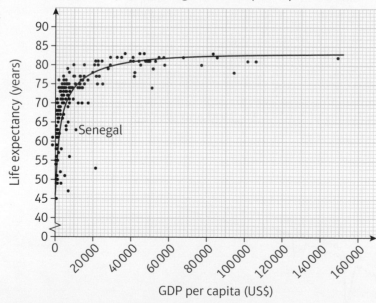

Source: Based on data from the World Bank

Table 1 shows the average life expectancy (for males and females) in the 20 countries from the sample used in Graph 3 that have the lowest GDP per capita.

Table 1 Life expectancy in countries with the lowest GDP per capita

Country	GDP per capita (nearest US$)	Life expectancy at birth, average (years)
Burkina Faso	673	56
Burundi	244	54
Central African Republic	470	49
Congo, Democratic Rep.	391	50
Eritrea	632	62
Ethiopia	470	63
Gambia, The	505	59
Guinea	487	56
Guinea-Bissau	559	54
Liberia	414	60
Madagascar	445	64
Malawi	270	55
Mali	642	55
Mozambique	580	50
Nepal	685	68
Niger	394	58
Rwanda	667	63
Sierra Leone	627	45
Togo	580	56
Uganda	670	59

Source: Based on data from the World Bank

Table 2 shows 20 more countries from the sample used in Graph 3. These countries have the highest GDP per capita. Again, the average life expectancy is shown for each country.

Table 2 Life expectancy in countries with the highest GDP per capita

Country	GDP per capita (nearest US$)	Life expectancy at birth, average (years)
Australia	67 512	82
Austria	48 348	81
Belgium	44 818	80
Bermuda	85 458	82
Canada	52 733	81
Denmark	57 636	80
Faeroe Islands	48 695	81
Finland	47 416	81
Ireland	48 391	81

continued

Country	GDP per capita (nearest US$)	Life expectancy at birth, average (years)
Japan	46 679	83
Kuwait	50 896	74
Liechtenstein	149 161	82
Luxembourg	106 023	81
Netherlands	49 128	81
Norway	101 564	81
Qatar	94 236	78
Singapore	54 578	82
Sweden	57 134	82
Switzerland	83 295	83
USA	51 457	79

Source: Based on data from the World Bank

Look at the data source

Using Graphs 1 and 2

1 What could have caused the dip in male life expectancy between 1911 and 1921?

2 Estimate the life expectancies of a 65-year-old male and a 65-year-old female in 2001.

3 Estimate their life expectancies at birth.

4 Comment on the differences between your answers to Questions 2 and 3. What factors do you think explain the differences?

Using Graph 3

5 A country has a GDP per capita that is $5000 greater than that of Senegal. Estimate the difference between the countries' life expectancies.

6 A country has a GDP per capita that is $10 000 greater than that of Senegal. Estimate the difference between the countries' life expectancies.

7 A journalist says, 'In wealthy countries, improvements in life expectancy have ceased to be related to improvements in wealth'. Use the shape of the graph to decide whether you agree with this statement.

Using Tables 1 and 2

Key point

A correlation coefficient measures the degree of correlation between two variables. The most well-known is the product moment correlation coefficient (PMCC), which has a value between -1 and 1:
$1 =$ perfect positive correlation, $-1 =$ perfect negative correlation and $0 =$ no correlation.

$$\text{PMCC} = \frac{n\Sigma xy - \Sigma x \Sigma y}{\sqrt{n\Sigma x^2 - (\Sigma x)^2}\sqrt{n\Sigma y^2 - (\Sigma y)^2}}$$

where $n =$ number of data points.

8 Calculate the product moment correlation coefficient for the data in Table 1.

Use $\Sigma x = 10\,405$, $\Sigma x^2 = 5\,749\,909$, $\Sigma y = 1136$, $\Sigma y^2 = 65\,144$, $\Sigma xy = 593\,649$ and $n = 20$.

Does this value for the PMCC mean that it is easy to predict life expectancy in countries with a low GDP? Why?

9 Calculate the product moment correlation coefficient for the data in Table 2.

Use $\Sigma x = 1\,345\,158$, $\Sigma x^2 = 104\,968\,343\,280$, $\Sigma y = 1615$, $\Sigma y^2 = 130\,487$, $\Sigma xy = 108\,787\,241$ and $n = 20$.

Does this value mean that it is easy to predict life expectancy in countries with a high GDP? Why?

Q8 hint

Draw a scatter diagram using the data to help you understand what the product moment correlation coefficient is showing.

Explore

Does a higher GDP cause a higher life expectancy?

2.2 Coping with risk

Data source

Much of what we do in life carries an element of risk. There are many situations in which we need to make decisions based on what might happen in the future, so we need a means of working out a probability associated with the risk to help us decide how to deal with it.

Financial risk

One way of reducing the impact of financial risk is to take out insurance against an unpleasant event that might happen or might not happen. In a typical insurance scheme, you pay a sum of money (a 'premium') relating to a particular risk in advance of a specific event. Insurance companies calculate the probability of the event happening and calculate how much the premium should be, based on that probability and on how much they are likely to need to pay out in the event of a 'claim'.

Typical insurance policies cover cars, houses, health, festivals and so on. Then, if the unpleasant event occurs (e.g. a theft, a fire, an accident, bad weather etc.), the insured person is compensated for the cost of putting things right. One way of deciding whether to take out insurance is to compare the cost of the insurance (a known amount) with the 'expected cost' of not taking out the insurance (the amount you would need to pay if something went wrong). For example, you might decide not to pay for an expensive insurance if you believe the likelihood of something unfavourable happening to an inexpensive item is very low.

Scenario 1

You have just bought an expensive new laptop. As you complete the purchase the shop assistant asks you if you would like to take out insurance against the laptop breaking down in the first three years. The insurance costs £30. The cost of a repair is £150.

Scenario 2

Garry, Selma and Sean are each buying a car. Legally, they must have 'third-party' insurance, which will compensate other people for any damage caused by them. To insure against repairs to their own car when they have caused the damage, drivers can also choose to take out 'comprehensive' insurance. This costs £300.

Scenario 3

A school is organising a fund-raising music festival in June. If it is a fine day, they expect 200 people to come and that each person attending will spend an average of £10. If it is raining, they expect about 50 people to come and that the average spend will be only about £5 per person. It is possible to take out insurance against rain.

The committee organising the festival is trying to decide between two insurance companies. Insurance company A charges a premium of £200 and will pay out £1500 if it rains. Insurance company B charges £50 and will pay out £1000 if it rains.

The committee has investigated the number of days it has rained in June over the past few years. The average is 3 rainy days.

Look at the data source

Key point

To assess risk, you can give it a numerical value based on the probability of an unfavourable event happening (P) and the consequences of that event happening (C):

$$\text{risk} = P \times C$$

In financial terms, C will be the cost of putting things right if the event happens.

Using Scenario 1

1 **a** The probability that the laptop will break down in the next three years is 1 in 20.

   i Compare the expected cost of taking out the insurance with that of not taking out the insurance.

   ii What is your decision on whether to take out the insurance? Explain your answer.

b Suppose the probability of the laptop breaking down was 25%.

  i Compare the expected cost of taking out the insurance with that of not taking out the insurance.

  ii Would your decision be the same as in part a? Explain your answer.

Using Scenario 2

2 To decide whether or not to take out comprehensive insurance, the drivers need to know the cost of repairs and the probability of needing to carry out repairs.

This table shows the information for Garry, Selma and Sean.

Driver	Probability of causing damage to their own car in one year	Average uninsured cost of a repair
Garry	0.1	£1100
Selma	0.3	£1100
Sean	0.5	£1100

For each driver

a calculate the total expected cost if they do not take out comprehensive insurance

b decide whether they should take out the comprehensive insurance, explaining your answer.

Using Scenario 3

3 a Work out the net expected income with each insurance company.

  b Which insurance company should the committee choose? Explain your answer.

Explore

What factors have an influence on car insurance premiums?

Q3a hint

You need to work out the expected income both for if it rains and for if it doesn't rain.

2.3 Marriages in England and Wales

Data source

Table 1 shows the total number of marriages each year in England and Wales from 1992 to 2012. Table 2 shows the number of marriages per quarter in England and Wales from 2009 to 2012.

Table 1 Number of marriages in England and Wales, 1992–2012

Year	Number of marriages	Year	Number of marriages
1992	311 564	2003	270 109
1993	299 197	2004	273 069
1994	291 069	2005	247 805
1995	283 012	2006	239 454
1996	278 975	2007	235 367
1997	272 536	2008	235 794
1998	267 303	2009	232 443
1999	263 515	2010	243 808
2000	267 961	2011	249 133
2001	249 227	2012	262 240
2002	255 596		

Source: Office for National Statistics

Table 2 Number of marriages per quarter in England and Wales, 2009–2012

Year	March quarter	June quarter	September quarter	December quarter
2009	26 808	64 990	97 045	43 600
2010	26 775	69 613	101 474	45 946
2011	25 581	72 267	103 680	47 605
2012	31 460	80 490	100 900	49 390

Source: Office for National Statistics

Look at the data source

Key point 1

A time series is a set of data points collected at intervals over a period of time.

A time series graph is a line graph plotted with time on the x-axis.

1 **a** Draw a time series graph to display the data in Table 1.

   b Describe the long-term trend.

   c Describe the short-term trend evident in the most recent five years of data.

2 **a** On a new grid, draw a time series graph to display the data in Table 2.

   b Describe and explain the shape of your graph.

Q2 hint

Time series data can exhibit long-term trends, short-term trends or seasonal fluctuations.

Key point 2

A moving average can be used to 'smooth out' data and display any overall trends more clearly. To calculate an n-point moving average, find the mean of n consecutive data values.

- Calculate the average of the first n data values.
- Repeat for the n data values starting from the second data value.
- Repeat for the n data values starting from the third data value.
- And so on.

Moving averages are plotted in the middle of the relevant time period.

3 **a** Calculate the four-point moving averages for the data in Table 2 and plot them on the same grid that you used in Question 2.

   b Do the moving averages reveal the same short-term trend as you described in Question 1c?

Q3a hint

The first four-point moving average should be plotted halfway between June 2009 and September 2009.

The second four-point average is the average of the data from June 2009 to March 2010.

Explore

Can you think of other data sets that might demonstrate seasonal trends?

When would 12-point moving averages be appropriate?

2.4 Social housing

Data source A

A property developer is putting forward a proposal to build the Weston housing estate. The local council insists that this new housing development is designed in a way that combines privately owned homes with rented social housing. This 'mixed tenure' approach is intended to break up areas of relative poverty and bring different social groups together. The developer wants to make as much profit as possible while still following the council's requirements.

The developer knows from market research that in this area most people want to buy three-bedroom houses with small gardens. The local council has asked the developer to also include smaller two-bedroom flats for the rented social housing.

The plot of land can be broken down into 210 small blocks of building space. Each flat takes up two blocks of space and each house uses five.

The council is also concerned about traffic problems. It has therefore restricted the total number of new dwellings to a maximum of 72. The council also insists that the number of flats is either equal to or greater than the number of houses.

The developer expects to make a profit of £20 000 on each flat and £60 000 on each house.

Technical literacy

A property developer arranges the construction of new properties or renovates existing buildings.

Data source B

The same property developer is asked to draw up plans for the nearby Easton estate, made up of two-bedroom flats for social housing and two-bedroom houses with gardens for private ownership.

The flats take up two blocks of building space, the houses take up four, and there are 300 blocks in total. Once again, the number of flats must be equal to or greater than the number of houses. There are no traffic restrictions for this estate, but the council is also asking that there be at least 20 houses.

The developer expects to make a profit of £20 000 on each flat and £50 000 on each house.

Look at the data sources

Using Data source A

1 Let x be the number of flats and y be the number of houses in the Weston estate.

   a Explain why the information about blocks of building space can be expressed as $2x + 5y \leqslant 210$.

   b Draw and label the graph of $2x + 5y = 210$.
      Use a grid with both axes going up to 110.

   c This line creates two regions. Shade the region that *doesn't* satisfy the inequality.

2 The council has restricted the total number of new dwellings.

   a Which of these inequalities properly expresses this constraint?
      $x + y > 72$ $72 + x < y$ $x + y \leqslant 72$ $72x \geqslant y$

   b Add this second inequality to your diagram and shade the region which does not satisfy this rule. Label your line.

3 The local council has also placed a restriction relating the number of flats to the number of houses.

   a Write this constraint as an inequality.

   b Add this information to your diagram and shade appropriately.

4 **a** Add $x \geqslant 0$ and $y \geqslant 0$ to your diagram and shade any regions you have excluded.

   b What do these inequalities mean in the real world?

Q1b hint

Where this line crosses the y-axis, the value of x is 0, so you can solve the equation $2 \times 0 + 5y = 210$ to find the y-intercept. Similarly, you can set $y = 0$ to find the x-intercept.

Q1c hint

If you're not sure which side of the line to shade, check one of the points. For example, the point $(100, 100)$ does *not* satisfy the inequality $2x + 5y \leqslant 210$ because $2 \times 100 + 5 \times 100 = 700$, so it should be in the *shaded* area.

Key point 1

The feasible region is the region that contains all of the possible pairs of values of x and y that would satisfy the constraints. For your diagram, the feasible region is the unshaded region.

5 Find the coordinates of the four vertices of your feasible region.

Q5 hint

If you have drawn your diagram accurately you might be able to read off the coordinates of the vertices. You can also work them out by solving simultaneous equations.

Key point 2

The objective function defines whatever it is that must be optimised. In this example, the developer wants to maximise profits, so the formula for the expected profit, P, that the developer expects to make can be used as the objective function.

6 Write down the objective function for the developer's profit.

7 a Copy and complete this table for the four vertices of your feasible region.

Number of flats, x	Number of houses, y	Profit, P (£)
0	0	0

b How many flats and how many houses should the developer build to maximise the profit?

Q7b hint

The optimal solution lies at one of the vertices of your feasible region.

8 a The council needs more social housing and decides to offer an additional £4000 bonus to this developer for each flat that is built. Write down the new objective function and re-calculate the profit at each vertex.

b Will this bonus change the developer's plans?
What would you advise the council to do?

Using Data source B

9 Design the Easton estate using linear programming.

   a Define your variables, write out your inequalities and draw a diagram showing the feasible region.

   b Write the new objective function and find the optimal solution for the developer planning the Easton estate.

Q9 hint

The process you have followed in Questions 1 to 8 is called linear programming.

Explore

What other real-life factors might lead to constraints when planning a housing development?

2.5 Population growth

Data source

Table 1 shows the estimated population of the United Kingdom between 1970 and 2010.

Table 1 Estimated population of the United Kingdom, 1970–2010

Decade	Initial population
1970s	55 632 200
1980s	56 329 700
1990s	57 237 500
2000s	58 886 100
2010s	62 759 500

Source: Office for National Statistics Overview of the UK Population

A researcher wants to use the data to predict the population in the 2020s. He has three possible models for population growth.

Model 1

The simplest model for population growth is an increasing arithmetic sequence:

$$a_{n+1} = a_n + d$$

Model 2

This model assumes that the population increases by a constant multiple, and uses an increasing geometric sequence:

$$b_1 = 55\,632\,200 \qquad \text{and} \qquad b_{n+1} = b_n \times 1.0306$$

Model 3

At any given time, a large part of the population will be in an age group that is unlikely to contribute to population growth. This model subtracts some of the population before multiplying:

$$c_1 = 55\,632\,200 \qquad \text{and} \qquad c_{n+1} = 1.69(c_n - 22\,310\,000)$$

Look at the data source

1   Draw a time series graph to display the population data in Table 1.

Subscript notation can be used to describe the terms of a sequence. For a sequence that starts 5, 7, 9, 11, ... , then $u_1 = 5$, $u_2 = 7$, $u_3 = 9$ and so on.

Sequence rules can be written out using

- a general formula for the nth term, u_n, for example $u_n = 2n + 3$, so you can calculate any term in the sequence immediately
- a term-to-term rule, where you give the first term, for example $u_1 = 5$, and the rule for getting from one term to the next, for example $u_{n+1} = u_n + 2$.

Key point 2

An arithmetic sequence can be given by the formula

$$u_{n+1} = u_n + d$$

where d is the common difference.

2   Use Model 1.

    a   Write down the values of the first term a_1 and the average common difference d.

    b   Work out the next four terms in this sequence, a_2, a_3, a_4 and a_5.

    c   Plot these five terms on your time series graph from Question 1.

    d   How well does this model match the observed values?

Key point 3

A geometric sequence is a sequence that increases or decreases by a fixed multiple. The number that you multiply each term by to find the next term is called the common ratio. The term-to-term rule is given by the formula

$$u_{n+1} = u_n r$$

where r is the common ratio.

3 Use Model 2.

  a What percentage increase in the population does the number 1.0306 represent?

  b Work out the first five terms in this sequence, b_1, b_2, b_3, b_4 and b_5.

  c Plot these five terms on your time series graph from Question 1.

  d Does this model match the observed values any more closely than Model 1?

4 Use Model 3.

  a Work out the first five terms in this sequence, c_1, c_2, c_3, c_4 and c_5.

  b Plot these five terms on your time series graph from Question 1.

  c Does this model match the observed values any more closely than Model 1 or Model 2?

5 **a** Calculate the predicted population in 2020 and 2030 using each of the models.

  b The Office for National Statistics has predicted a UK population of 67 359 600 in 2020.
Which of the three models matches this prediction most closely?

6 Comment on the three models.

Explore

What might you do to create a better, fourth model?

Why might population projections influence government policies on health, welfare and housing?

Q5a hint

The value for 2015 is the 6th term in the sequence.

Q6 hint

Which model do you think is the most successful at matching the observed data?

Which do you think makes the most realistic assumptions about human behaviour?

3 Sport

In this unit you will work on projects involving various sports, including football, tennis, golf and athletics.

You will:

- Look for relationships between performance data and earnings
- Investigate the improvements in 100-metre sprint times
- Identify relationships in football club data
- Calculate probabilities in tennis.

Prior knowledge

Before beginning this unit make sure you can:

- Draw scatter diagrams
- Calculate percentages
- Find quartiles and the interquartile range for a data set.

3 Sport

3.1 Golf

Data source A

Tables 1 and 2 show the top 20 money earners in a women's lower tier Iberian professional golf tour and in a men's professional golf tour in the same region during 2015. The tables also show how many tournaments each player competed in, how many tournaments they won and how many times they finished in the top ten. For example, Alessia Rossi won three tournaments and had a total of 17 top-ten finishes.

Table 1 Top 20 earners in women's Iberian golf tour, 2015 season

Rank	Name	Winnings (€)	Events	Wins	Top tens
1	Alessia Rossi	57 571	25	3	17
2	Sarah Turner	49 767	20	3	15
3	Raquel Pereira	46 722	23	3	16
4	Ji-Woo Kweon	42 701	18	2	13
5	Francesca Marinos	32 436	22	1	13
6	Aurora Bianchi	31 435	21	1	11
7	Beatriz Soares	25 948	23	2	7

Rank	Name	Winnings (€)	Events	Wins	Top tens
8	Margherita Rojas	24 052	16	2	5
9	Chiara Harrison	23 986	24	0	9
10	Margarida Rosa	23 403	28	0	9
11	Carmen Pena	22 185	21	0	9
12	Patrícia Morais	20 847	22	1	7
13	Jie Xue	20 351	23	0	5
14	Maria Vega	20 025	21	2	4
15	Yasmine Darzi	19 277	21	0	9
16	Elisa Marino	18 166	20	2	6
17	Marii Matos	17 472	23	0	5
18	Rebecca Kirsch	17 260	25	0	7
19	Sofia Laurent	16 706	24	0	5
20	Adriana Costa	16 073	23	0	7

Table 2 Top 20 earners in men's Iberian golf tour, 2015 season

Rank	Name	Winnings (€)	Events	Wins	Top tens
1	Javier Martinez	85 216	14	3	11
2	Georg Fischer	66 486	18	2	8
3	Joao Martins	61 882	18	0	11
4	Iker Valdez	58 849	24	3	9
5	Daniel Russo	52 962	13	0	10
6	Lucas Davies	50 276	25	2	5
7	Hélder Barros	50 256	24	2	5
8	Yong Chen	49 721	23	0	10
9	Riad Kader	48 669	21	1	11
10	Mateo Ruiz	48 322	16	2	4
11	Bruno Gerste	47 042	24	0	8
12	Rafael Spiros	45 543	14	1	7
13	Nadir Zabat	44 036	14	1	8
14	Samuel Bauer	42 738	25	2	3
15	Nuno Oliveira	41 869	16	1	7
16	Fernando Leon	38 765	12	1	5
17	Lorenzo Russo	36 253	22	1	8
18	Tomas Soares	36 035	26	1	6
19	Olivier Laurent	35 609	23	1	4
20	Michael Reis	34 704	24	0	5

Data source B

Table 3 shows a random sample of 20 female Central European golfers from the 2015 season and how much money they won. All female golfers who competed in at least 10 tournaments are included in the population from which the sample is drawn.

Table 3 Random sample of Iberian female golfers, 2015 season

Number	Rank	Name	Winnings (€)	Events
1	11	Carmen Pena	22 185	21
2	25	Simone Roux	9832	20
3	36	Ines Sousa	8509	26
4	45	Gloria Tan	7369	24
5	46	Paola Esposito	7290	18
6	53	Kristin Seuss	6519	27
7	57	Francine Dubois	6625	24
8	59	Catarina Vieira	6389	16
9	62	Olivia Greco	6111	28
10	69	Sara Gonçalves	5762	26
11	74	Joana Martins	5135	21
12	75	Martina Novak	5119	22
13	77	Valeria Ricci	5004	19
14	91	Liah Guedes	4670	16
15	104	Caterina Hofer	4211	15
16	109	Laura Freitas	4089	18
17	112	Carolina Valente	3554	17
18	133	Valentina Szabo	2653	13
19	155	Alba Lopez	1360	9
20	156	Taha Medina	1125	10

Look at the data sources

Using Data source A

1 **a** Record the minimum and maximum winnings, and work out the lower quartile, median and upper quartile, in each data set.

    b Work out the interquartile range for the winnings in each set of data.

    c Draw two box plots, on the same scale, to show these data sets.

Key point 1

A box plot displays the maximum and minimum values, the median and the quartiles of a data set.

Key point 2

The lower quartile (LQ), upper quartile (UQ) and interquartile range (IQR) can be used to identify outliers.

Any data value lower than $LQ - 1.5 \times IQR$ or higher than $UQ + 1.5 \times IQR$ is an outlier.

2　**a**　Comment on whether either of these sets of data has outliers.

　　b　Explain why it is not a good idea to work out the mean and standard deviation of these data sets.

3　Compare the distributions.

Key point 3

Spearman's rank correlation coefficient has a value between -1 and 1:

1 = perfect positive correlation, -1 = perfect negative correlation and 0 = no correlation.

To calculate Spearman's rank correlation coefficient:

- Use the first variable, x, to put the data values in order and give them a rank number, x_i. If two or more data values are the same, their rank number is the average of those ranks. For example, if the 2nd and 3rd data values are the same, $x_i = 2.5$ for both.
- Repeat for the second variable, y.
- For each data item, calculate $d_i = x_i - y_i$ and then calculate d_i^2.
- Calculate the correlation coefficient using the formula

$$\text{Spearman's rank correlation coefficient} = 1 - \frac{6\Sigma d_i^2}{n(n^2 - 1)}$$

where n is the number of data values.

Example

x	y	x_i	y_i	d_i	d_i^2
4	26	1	2	−1	1
7	13	2	1	1	1
11	37	3.5	3	0.5	0.25
11	76	3.5	5	−1.5	2.25
19	39	5	4	1	1
25	97	6	6	0	0
				Total	5.5

$$\text{Spearman's rank correlation coefficient} = 1 - \frac{6 \times 5.5}{6 \times (36 - 1)}$$
$$= 0.8429 \text{ (to 4 d.p.)}$$

Key point 4

If there is a causal relationship, the x-axis takes the independent or fixed variable, and the y-value is the one caused by x.

4 You are going to draw a scatter diagram to show the relationship between money earned and the number of top-ten finishes for the female golfers.

   a Which variable should be plotted on the x-axis and which should be plotted on the y-axis?

   b Draw a scatter diagram for this data.

   c Calculate the Spearman rank correlation coefficient for the data.

   d Comment on the strength of the correlation shown in your diagram.

Q4c hint

You can use a spreadsheet to make the calculation easier.

Using Data source B

5 Gina wants to work out how much a golfer can expect to earn on average on the Iberian tour.

   a Explain why the data set in Table 3 is better to use than the one in Table 1.

   b Work out the minimum, lower quartile, median, upper quartile, maximum and interquartile range for this data set.

   c Does the data set contain outliers? If yes, which golfers are the outliers?

Explore

Research some real golf tournaments. How much would it cost to travel to 20 different tournaments per year? How much does a golfer need to earn to cover their costs and make a living?

3.2 Athletics

Data source

Table 1 shows the fastest 100-metre sprint times run each year since 1980.

Table 1 Fastest 100-metre sprint times in year, 1980–2015

Year	Men's best time	Women's best time	Year	Men's best time	Women's best time
1980	10.02	11.02	1998	9.86	10.65
1981	10.00	10.99	1999	9.79	10.70
1982	10.00	10.95	2000	9.86	10.78
1983	9.93	10.79	2001	9.82	10.82
1984	9.96	10.76	2002	9.89	10.83
1985	9.98	10.98	2003	9.93	10.86
1986	10.00	10.91	2004	9.85	10.77
1987	9.93	10.86	2005	9.77	10.84
1988	9.92	10.49	2006	9.77	10.82
1989	9.94	10.78	2007	9.74	10.89
1990	9.96	10.78	2008	9.69	10.78
1991	9.86	10.79	2009	9.58	10.64
1992	9.93	10.80	2010	9.78	10.78
1993	9.87	10.82	2011	9.76	10.70
1994	9.85	10.77	2012	9.63	10.70
1995	9.91	10.84	2013	9.77	10.71
1996	9.84	10.74	2014	9.77	10.80
1997	9.86	10.76	2015	9.74	10.74

Source: Based on information from www.iaaf.org

Look at the data source

Key point

A regression line is the line that matches the pattern of data as closely as possible. The least squares regression line is the one that has the smallest possible value for the sum of the squares of the y-distances of all the data points from the line.

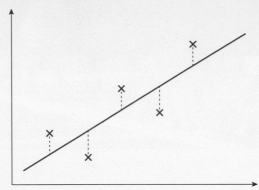

The equation of the least squares regression line is given by

$$y = a + bx$$

where $b = \dfrac{n\Sigma xy - \Sigma x \Sigma y}{n\Sigma x^2 - (\Sigma x)^2}$

$a = \bar{y} - b\bar{x}$

n = number of data points

$\bar{x}$ = mean of x-values

$\bar{y}$ = mean of y-values.

1　**a**　Plot a scatter diagram for the men's sprint times with the year on the x-axis and the sprint time on the y-axis.

　　b　Work out $\bar{x}$ and $\bar{y}$.

　　c　Calculate the equation of the least squares regression line.
　　　Use $\Sigma x = 71\,910$, $\Sigma x^2 = 143\,644\,110$, $\Sigma y = 354.76$,
　　　$\Sigma xy = 708\,599.08$ and $n = 36$.

　　d　Plot the regression line on your scatter diagram.

　　e　According to the regression line, what is happening to the 100-metre times each year?

　　f　Use the model to predict the fastest men's 100-metre time in the year 2100. Is this time realistic?

　　g　What will happen in the year 3123?

2 a Calculate three-point moving averages for both the men's data and the women's data, using the data points for every fifth year.

Year	Men	Moving average	Women	Moving average
1980	10.02		11.02	
1985	9.98		10.98	
1990	9.96		10.78	
1995	9.91		10.84	
2000	9.86		10.78	
2005	9.77		10.84	
2010	9.78		10.78	
2015	9.74		10.74	

b Comment on the trends in the men's and women's data.

Q2a hint

There are no three-point averages for 1980 or 2015 because there are not three data points centred on these years.

The three-point average for 1985 is the average of the 1980, 1985 and 1990 values.

3 a Calculate the product moment correlation coefficient (PMCC) for the women's data.
Use $\Sigma x = 71\,910$, $\Sigma x^2 - 143\,644\,110$,
$\Sigma y = 388.64$, $\Sigma y^2 = 4195.95$ and $\Sigma xy = 776\,412.35$

b The PMCC of the men's data is -0.868.
Comment on the differences in correlation between the men's data and the women's data.

c Graham says that because the correlation between year and 100-metre time is so strong for the men's data, it is clear that the year is causing the decrease in 100-metre times. Explain why Graham might not be correct.

Q3a hint

$$PMCC = \frac{n\Sigma xy - \Sigma x\Sigma y}{\sqrt{n\Sigma x^2 - (\Sigma x)^2}\sqrt{n\Sigma y^2 - (\Sigma y)^2}}$$

where n = number of data points.

Explore

Look at records in other athletic events – do they show the same improvement over time?

What about other sports? Does the fastest lap at Silverstone in the F1 Grand Prix improve every year?

3.3 Football

Data source

Table 1 shows information about all the Barclays Premier League clubs for the 2013–14 season.

Table 1 Data on clubs' league performance, match attendance and ground information, 2013–14 season.

Team	League position	Annual turnover (£ million)	Average weekly salary (£)	Capacity of ground*	Average attendance	Price of cup of tea (£)	Cheapest ticket (£)
Arsenal	4	304	77 963	60 338	59 487	2.00	26.00
Aston Villa	15	117	34 815	42 682	36 081	2.10	23.00
Cardiff City	20	83	25 943	27 815	27 430	1.70	20.00
Chelsea	3	324	83 713	41 798	41 482	2.20	36.00
Crystal Palace	11	90	19 205	26 255	24 114	2.20	25.00
Everton	5	121	31 447	39 571	37 732	2.20	32.00
Fulham	19	91	29 761	25 700	24 977	1.90	25.00
Hull City	16	84	20 215	25 400	24 117	2.00	22.00
Liverpool	2	256	67 486	45 276	44 671	2.40	38.00
Manchester City	1	347	96 445	47 405	47 080	1.80	20.00
Manchester United	7	433	89 988	75 731	75 207	2.50	31.00
Newcastle United	10	130	32 223	52 405	50 395	2.20	15.00
Norwich City	18	94	22 305	27 244	26 805	2.10	20.00

*at the beginning of the season

Team	League position	Annual turnover (£ million)	Average weekly salary (£)	Capacity of ground*	Average attendance	Price of cup of tea (£)	Cheapest ticket (£)
Southampton	8	106	27 081	32 589	30 212	2.00	30.00
Stoke City	9	98	25 938	27 740	26 137	2.10	25.00
Sunderland	14	104	31 078	48 707	41 090	2.20	25.00
Swansea City	12	98	31 015	20 750	20 407	2.00	35.00
Tottenham Hotspur	6	181	54 077	36 284	35 808	2.00	32.00
West Bromwich Albion	17	87	29 359	26 445	25 194	1.80	25.00
West Ham	13	115	27 394	35 016	34 197	2.00	37.00

Sources: Based on information from www.theguardian.com, www.worldfootball.net, www.premierleague.com, www.mirror.co.uk and www.fsf.org.uk

*at the beginning of the season

Look at the data source

1 a Calculate the minimum, lower quartile (LQ), median, upper quartile (UQ), maximum and interquartile range (IQR) of the salary data.

b Comment on whether the data has outliers.

c Draw a box plot to represent the data.

Key point

Data is skewed when it is not symmetrical about the median.

For symmetric data, median ≈ mean.

For data that is skewed right (or positively skewed), mean > median, that is, the mean is to the right of the median.

For data that is skewed left (or negatively skewed), mean < median, that is, the mean is to the left of the median.

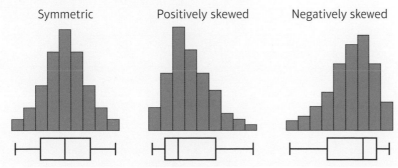

Symmetric — Positively skewed — Negatively skewed

Q1b hint

Any data value lower than LQ − 1.5 × IQR or higher than UQ + 1.5 × IQR is an outlier.

Q1c hint

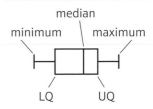

2 a Calculate the minimum, lower quartile, median, upper quartile, maximum and interquartile range of the annual turnover data.

b Comment on whether the data has outliers.

c Draw a box plot to represent the data.

d Does each data set show positive skew or negative skew?

e Theresa says that the Premier League is an unfair league because a few teams possess most of the money. Do your results support Theresa's statement?

3 **a** Calculate Spearman's rank correlation coefficient for each of the following pairs of data:

  i League position and average weekly salary

  ii League position and annual turnover

  iii League position and cheapest ticket.

b Comment on the values that you have calculated.

c The PMCCs for the same pairs of data are:

League position and average weekly salary	−0.743
League position and annual turnover	−0.722
League position and cheapest ticket	−0.435

  i Why are these values negative?

  ii Comment on the similarities and differences between the PMCC and Spearman values.

4 **a** What other factors might affect a team's league position?

b Ramana says teams should just generate more money (have a higher turnover) if they want to improve their league position. Is Ramana correct?

Explore

Investigate whether any other factors in the table show a correlation with league position.

Do other football leagues show similar relationships?

Do other sports show similar relationships?

Q3a hint

Rank the league positions from best (rank 1) to worst (rank 20).

Rank the salaries, turnover and cheapest tickets from highest (rank 1) to lowest (rank 20).

Spearman's rank correlation coefficient
$$= 1 - \frac{6\Sigma d_i^2}{n(n^2 - 1)}$$

3.4 Tennis

Data source A

In men's tennis, Novak Djokovic, Roger Federer, Andy Murray and Rafael Nadal are sometimes referred to as 'The Big Four'. This means a lot of people think of them as being the four dominant players of this era, from the early 2000s to the mid-2010s.

Tables 1 and 2 show their records against one another, in terms of matches and sets. This includes all matches that were awarded a victory due to 'walkovers', or the opponent not playing for any reason.

Table 1 Matches won/lost by 'The Big Four' playing against each other

		Matches lost			
		Djokovic	**Federer**	**Murray**	**Nadal**
Matches won	**Djokovic**		22	20	22
	Federer	21		14	11
	Murray	9	11		6
	Nadal	23	23	15	

Source: Based on data from www.atpworldtour.com

Table 2 Sets won/lost by 'The Big Four' playing against each other

		Sets lost			
		Djokovic	Federer	Murray	Nadal
Sets won	Djokovic		59	49	60
	Federer	65		40	41
	Murray	32	27		21
	Nadal	61	65	40	

Source: Based on data from www.atpworldtour.com

Data source B

Tables 3 and 4 contain data relating to matches between Andy Murray and Rafael Nadal. Table 3 shows the outcomes of the first set and of the match. Table 4 shows who goes on to win the next set after Andy Murray has won a set and similarly after Rafael Nadal has won a set.

Table 3 First set and match results for Rafael Nadal against Andy Murray

Match number	First set	Match
1	Murray	Murray
2	Nadal	Nadal
3	Murray	Nadal
4	Nadal	Murray
5	Nadal	Nadal
6	Murray	Nadal
7	Nadal	Nadal
8	Nadal	Nadal
9	Nadal	Nadal
10	Murray	Murray
11	Nadal	Nadal
12	Murray	Murray
13	Nadal	Nadal
14	Nadal	Nadal
15	Murray	Murray
16	Murray	Murray
17	Nadal	Nadal
18	Nadal	Nadal
19	Nadal	Nadal
20	Nadal	Nadal
21	Murray	Nadal

Source: Based on data from www.atpworldtour.com

Table 4 Next-set outcomes between Andy Murray and Rafael Nadal related to who won the previous set

Murray wins a set		Nadal wins a set	
Murray wins next set	Nadal wins next set	Murray wins next set	Nadal wins next set
6	9	7	18

Source: Based on data from www.atpworldtour.com

Data source C

Table 5 summarises the career results so far of 'The Big Four' together with those of other leading players of the same era.

Table 5 Career results for leading tennis players

Name	Age	Turned professional	Tournament wins (including Grand Slam events)	Grand Slam wins	Money won ($)
Novak Djokovic	28	2003	57	10	88 444 918
Roger Federer	34	1998	88	17	95 198 691
Andy Murray	28	2005	35	2	40 320 055
Rafael Nadal	29	2001	67	14	74 561 205
Juan Martin del Potro	27	2005	18	1	15 369 422
David Ferrer	33	2000	26	0	27 621 396
Richard Gasquet	29	2002	12	0	13 019 661
Lleyton Hewitt	34	1998	30	2	20 717 156
Andy Roddick	33	2000	32	1	20 640 030
Marat Safin	35	1997	15	2	14 373 291
Jo-Wilfried Tsonga	30	2004	12	0	16 389 610
Stan Wawrinka	30	2002	11	2	19 559 208

Source: Based on data from www.atpworldtour.com

Look at the data sources

Using Data source A

Where necessary give your answers to 3 significant figures.

1 Novak Djokovic has won 22 matches against Roger Federer.

 a How many matches has Roger Federer won against Novak Djokovic?

 b What percentage of the total matches played between the two did Roger Federer win?

 c Copy and complete the table to show what percentage of matches each player won and lost against the other.

		Matches lost (%)			
		Djokovic	Federer	Murray	Nadal
Matches won (%)	Djokovic				
	Federer				
	Murray				
	Nadal				

 d Create a similar table to show what percentage of sets each player won and lost against the other.

 e How would you rank the four players based on their performance against one another?

2 Use the data from Question 1 to answer this question. Assume that past results can predict future performance and that set wins are independent.

 a Explain what 'set wins are independent' means.

 b In a match between Murray and Nadal, what is:

   i P(Murray wins a set)

   ii P(Nadal wins a set)?

 c Draw a tree diagram to show the possible outcomes for the first two sets of a match.

 d Extend your tree diagram to show the possible outcomes for a best-of-three-set match.

 e What is the probability that Murray would beat Nadal in a three-set match?

Q2c hint

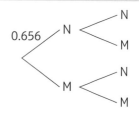

Q2d hint

One player has to win two sets before the match is over.

Using Data source B

3 a Copy and complete this table.

	Murray win percentage	Nadal win percentage
Overall sets	34.4%	65.6%
First set of the match		
Set following a Murray set win		
Set following a Nadal set win		

b Some people think that 'momentum' plays a role in tennis: the result of the previous set influences the result of the next set. Does the data support this?

c Use the data from the table from part a to create a new tree diagram for the possible outcomes of a three-set match.

d Using the tree diagram from part c, what is the probability that Murray would beat Nadal in a three-set match?

e Compare your answers to Questions 2e and 3d. Comment on your results.

Key point

For events A and B:

$A \cap B$ $A \cup B$ A' (not A)

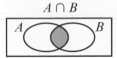 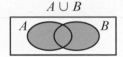 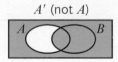

4 a Complete this Venn diagram to represent the first set and match information in the Murray–Nadal rivalry.

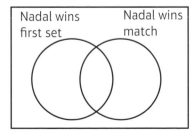

Nadal wins first set   Nadal wins match

b Use your answer to Question 1c to give the probability that Nadal beats Murray in a match.

c Given that Nadal wins the first set, what is the probability he also wins the match?

d Comment on whether who wins the match is independent of who wins the first set.

Using Data source C

5 a Draw a scatter diagram showing tournament wins and money won.

b Calculate the equation of the least squares regression line. Use $\Sigma x = 403$, $\Sigma x^2 = 20\,265$, $\Sigma y = 446.22$ million (to 2 d.p), $\Sigma xy = 22\,866.11$ million (to 2 d.p.) and $n = 12$. Draw the line on your scatter diagram.

c Calculate the product moment correlation coefficient (PMCC).

d Comment on the strength and validity of your model.

e John McEnroe turned professional in 1978 and won 77 tournaments and \$12 552 132.
Explain why the model does not more accurately predict his winnings.

f Use the model to predict how much John McEnroe would have won if he had played in this era.

Q5b hint

The equation of the least squares regression line is given by
$$y = a + bx$$
where $b = \dfrac{n\Sigma xy - \Sigma x \Sigma y}{n\Sigma x^2 - (\Sigma x)^2}$
$a = \bar{y} - b\bar{x}$

n = number of data points
$\bar{x}$ = mean of x-values
$\bar{y}$ = mean of y-values.

Q5c hint

$$\text{PMCC} = \dfrac{n\Sigma xy - \Sigma x \Sigma y}{\sqrt{n\Sigma x^2 - (\Sigma x)^2}\sqrt{n\Sigma y^2 - (\Sigma y)^2}}$$
where n = number of data points.

Explore

In Question 5f you predicted how much John McEnroe would have won if he had played in this era. How does McEnroe compare to professional tennis players from other eras?

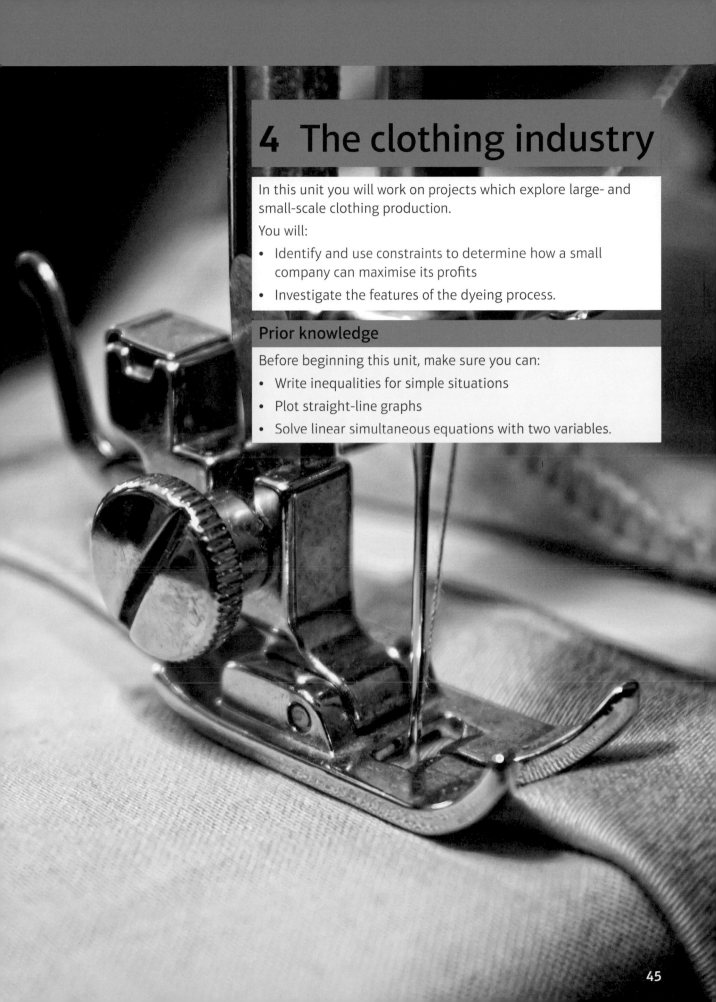

4 The clothing industry

In this unit you will work on projects which explore large- and small-scale clothing production.

You will:

- Identify and use constraints to determine how a small company can maximise its profits
- Investigate the features of the dyeing process.

Prior knowledge

Before beginning this unit, make sure you can:

- Write inequalities for simple situations
- Plot straight-line graphs
- Solve linear simultaneous equations with two variables.

4 The clothing industry

4.1 Manufacturing baby clothes

Data source

To prepare for the 150th anniversary of the birth of Beatrix Potter in 2016, a small clothing company made limited-edition babies' dungarees and jackets with Beatrix Potter characters on them.

The company wants to maximise the money they make from sales. They assume that they will sell all the items.

The dungarees and jackets are made from the same yellow denim.

The dungarees require 4000 cm² of denim and the jackets use 5000 cm² of denim.

They want to include 2 character buttons on each item of clothing.

The dungarees take 90 minutes to make and the jackets take 1 hour to make.

The staff can devote 100 hours of labour in total to make these limited-edition dungarees and jackets.

In total they have 35 m² of denim and 170 buttons to use.

The company will sell each pair of dungarees for £25 and each jacket for £25.

Look at the data source

1   The company wants to work out how many pairs of dungarees and how many jackets to make. Define the variables involved in this problem.

2   Write an inequality to represent the constraints on
   a amount of fabric   **b** labour   **c** character buttons.

3   **a** Represent the inequality for the amount of fabric on a graph.
   b Shade the region which does not satisfy the inequality.

4   **a** Represent the inequalities for labour and character buttons from Question 2 on your graph.
   b Shade the regions which do not satisfy the inequalities.

5   Draw lines on your graph to show that the numbers of dungarees and jackets the company produces cannot be negative.

6   **a** State the company's business aim.
   b Write a formula for the objective function, P.

Key point

In Unit 2, you used the values at the four vertices of the feasible region to find the optimal value. Here is another way.
Choose a sensible value of P so you can plot a graph of the objective line.

- If this line does not cross the feasible region, choose a different value.
- Imagine the line sliding across the feasible region away from the origin (you could use a transparent ruler on your graph, keeping it parallel to the objective line).
- The maximum income occurs when the objective line reaches the furthest point from the origin within the feasible region.

7   Using a first estimate of $P = 1000$, plot an objective line on your graph.
   a Use your graph to estimate the numbers of dungarees and jackets the company can make to generate the maximum income.
   b What is the maximum income?

8   Solve the simultaneous equations algebraically to check your estimate from the graph.

9   The company didn't need to use all the buttons. How many were left over?

Explore

Are these clothes sufficiently profitable for the company? What other things should you take into consideration?

Q1 hint

Don't forget to make sure your units are the same.

Q3 hint

Work out the x- and y-intercept for each line to ensure your axes are long enough.

Q8 hint

Which two constraints do you need to consider?

Each item of clothing needs to be finished, and the total must lie within the feasible region.

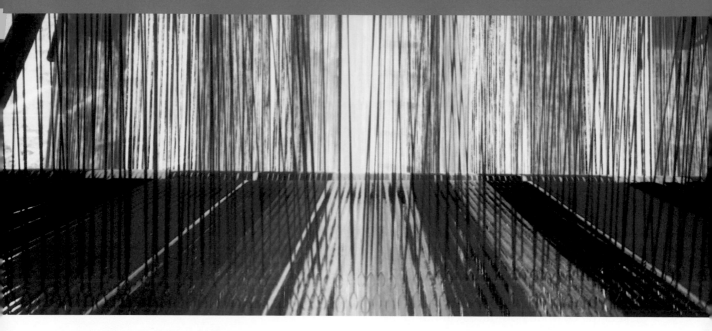

4.2 Dyeing fabric

Data source A

One way to dye fabric is to use a vat dye. The fabric is put in a container with dye, water and salt. The depth of the resulting colour changes linearly according to the concentration of dye in the liquid. The concentration of the dye is measured in grams per litre (g/l) and a scale of 0–10 can be used to identify the depth of colour achieved for a given concentration of dye. At 0 the colour can't be detected, while 1 is a weak colour and 10 is the strongest colour possible with that type of dye.

The depth of colour achieved with a particular dye concentration varies depending on the type of fabric, especially if it contains polyester or other synthetic fibres.

When one type of red dye is used, the human eye can start to pick up the red colour when the dye concentration is approximately 3.3 g/l. The maximum colour depth for this red dye on cotton is achieved using a concentration of 20 g/l.

Table 1 shows the depths of colour on cotton and on polycotton for other concentrations.

Table 1 Depth of red colour on cotton and polycotton depending on dye concentration

		Colour depth on cotton	Colour depth on polycotton
Concentration	**5 g/l**	1	0
	20 g/l	10	8.5

Data source B

Colour fastness is the extent to which a colour resists fading or running once a fabric has been dyed.

Temperature affects the colour fastness of the dye, and most dyes have an optimum temperature. Adding sodium carbonate (Na_2CO_3) to the water can also help obtain a higher level of colour fastness.

Graph 1 shows the colour fastness obtained when different temperatures are used when dyeing 1 kg of cotton fabric with a concentration of 15 g/l of red dye.

Graph 1 Variation of colour fastness with dyeing temperature

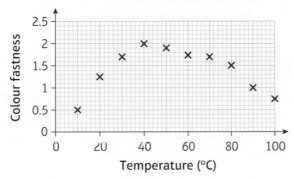

In order to use the dye in the most efficient way, one vat of dye can be used multiple times. However, the colour depth of subsequent batches of fabric will change because the concentration of the remaining dye in a used vat depends on how much dye was taken up by the previous batch. This in turn depends on the dyeing temperature.

This recurrence relationship shows how the concentration, decreases when the dye has been used previously:

$$c_{n+1} = c_n - 10t^{-1}$$

where c is the dye concentration in g/l

   n is the number of times the dye has been used previously

   t is the dyeing temperature in °C, within a normal range.

Look at the data sources

Using Data source A

1 **a** Plot a straight-line graph for the concentration and colour depth on cotton fabric.
Plot concentration on the x-axis and colour depth on the y-axis.

b Write the equation of your line in the form $y = mx + c$.

c Why does the equation only make sense for values of c between 3.3 g/l and 20 g/l?

2 The equation of the line for polycotton is approximately $y = 0.6x - 3.5$. Use the equations to explain why it is not possible to dye cotton and polycotton to the same colour depth using the same vat of dye.

Using Data source B

3 What is the optimum temperature for dyeing cotton using the red dye?

4 **a** A company wants to dye a batch of cotton fabric red with depth of colour 8 and using the optimum temperature for the red dye. What will be the depth of colour of a second batch of cotton that is dyed using the remaining dye? Give your answer to the nearest whole number.

b How many batches of cotton can be dyed at the optimum temperature before there is a visible change in the colour? Explain your answer.

Q4a hint

What is the initial concentration for achieving a colour depth of 8?

Explore

Look at Graph 1. What type of curve can you draw to model the data points?

5 Finance

In this unit you will work on projects involving income tax, life insurance, loans, savings and investments.

You will:

- Compare income tax in different countries
- Explore finance options
- Analyse salary and house price data
- Calculate the cost of life insurance.

Prior knowledge

Before beginning this unit, make sure you can:

- Convert between different units of currency
- Interpret and use formulae
- Use percentages
- Draw a cumulative frequency graph.

5 Finance

5.1 Income tax

Data source

Tables 1 to 3 show the tax rates in the UK and in two other countries in the EU. The amounts shown are total annual earnings.

Table 1 UK tax rates

Tax threshold	0% on any annual earnings up to £10 000
Basic tax rate	20% on annual earnings above the tax threshold and up to £41 865
Higher tax rate	40% on annual earnings from £41 866 to £160 000
Additional tax rate	45% on annual earnings above £160 000

Source: Adapted from www.gov.uk

Table 2 Tax rates in Country A

Tax threshold	0% on any annual earnings up to €7664
Band 1	15% on annual earnings above the tax threshold and up to €52 153
Band 2	42% on annual earnings from €52 154 to €250 000
Band 3	45% on annual earnings above €250 000

Table 3 Tax rates in Country B

Tax threshold	0% on any annual earnings up to €9690
Band 1	14% on annual earnings above the tax threshold and up to €26 764
Band 2	30% on annual earnings from €26 765 to €71 754
Band 3	41% on annual earnings from €71 755 to €151 956
Band 4	45% on annual earnings above €151 956

Look at the data source

1   Use the exchange rate €1 = £0.74 to convert the tax thresholds in Countries A and B into pounds.
    Round your answers to the nearest pound if necessary.

2   Helen lives in the UK and earns £72 000 annually. She creates this formula to work out how much tax she pays:

$$T = 0.4(x - 41\,865) + 0.2(41\,865 - 10\,000) + 0(10\,000 - 0)$$

where T is the tax paid and x is her annual salary.

   a   The first part of the formula is $0.4(x - 41\,865)$.
            Explain where the 0.4 comes from.

   b   Explain why the last part of the formula, $0(10\,000 - 0)$, is unnecessary.

   c   Work out the annual tax that Helen pays.

   Robert earns £24 500 annually.

   d   Explain why Robert cannot use Helen's formula to work out his tax bill.

   e   Write a formula that Robert can use.

   f   Work out how much tax Robert pays.

3   Helen and Robert are both considering moving abroad.
    Assume that their salaries do not change.

   a   Write a formula for how much tax each of them would pay in
      i Country A
     ii Country B.

   b   Work out how much tax each of them would pay in
      i Country A
     ii Country B.

Q3 hint

Work in pounds, using your answers to Question 1.

4   Compare the amounts of tax paid by a person earning Helen's or Robert's salary in the UK, Country A and Country B.

5   Where would a person earning an annual salary of £1 000 000 pay
   a   the least tax
   b   the most tax?

Q5 hint

Use the tables to answer this question without making any formal calculations.

Explore

Find two countries in the EU with similar tax rates to Countries A and B. How do these tax rates compare with the rates in other countries in the world?

Do you think the assumption in Question 3 is reasonable?

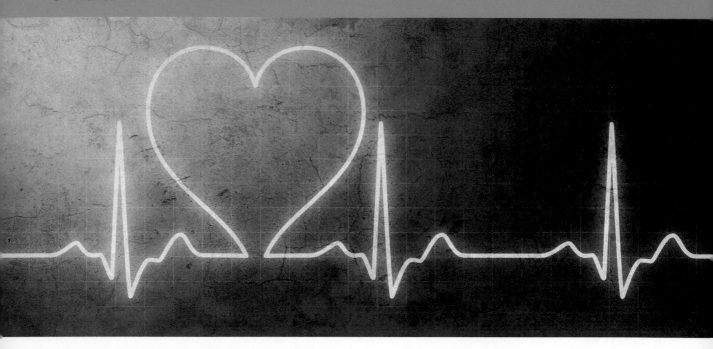

5.2 Life insurance

Alison Young, aged 34, and Graham Young, aged 37, are married and in good health. They are interested in buying life insurance to allow the remaining partner to stay in the same house and not worry about money for a few years should one of them die.

They have investigated different options, but haven't made a final decision.

There are many different types of life insurance, but two types are most common: term life insurance and whole life insurance. Another option Alison and Graham are considering increasing their savings.

Term life insurance

Term life insurance is life insurance that covers the insured person for a fixed period of time (or 'term'), usually 10 years, 20 years or 30 years. If the person dies during the term of the insurance a pay-out is made. If the person dies after this time, there is no pay-out and the insurance company keeps the premiums.

Table 1 shows the monthly premiums for £200 000 worth of life insurance cover for different terms, for someone who is in good health. People who can expect to pay more than the monthly premiums in Table 1 include those who

- smoke
- are overweight
- have a history of disease in their family
- participate in extreme sports
- have a hazardous occupation.

Alison and Graham do not fall into any of these categories.

Table 1 Monthly premium for £200 000 worth of term life insurance

Age	Male			Female		
	10 year	20 year	30 year	10 year	20 year	30 year
25–29	£13.23	£19.76	£32.42	£11.22	£16.83	£28.94
30–34	£15.76	£22.02	£37.01	£13.01	£18.96	£33.65
35–39	£19.45	£28.64	£43.98	£17.68	£25.33	£39.60
40–44	£23.87	£31.23	£48.02	£21.26	£28.99	£43.69

Whole life insurance

Whole life insurance provides coverage for the entirety of the insured person's life, provided they continue to pay the monthly premiums. Owing to the fact that everyone will die eventually, these monthly premiums are typically higher than for term life insurance.

Savings

Another option would be for the Youngs to put extra money into their savings each month. An advantage of this method is that the money is always available and could be used to pay other future expenses instead. A disadvantage is that there might be less money in their savings than life insurance would have paid out. Alison and Graham recently spoke to a financial advisor who said they could expect an annual return of 5% on their savings investment.

Look at the data source

1   Alison and Graham are interested in term life insurance for the next 10 years.

   a   What is the monthly premium for
      i   Alison
      ii Graham?

   b   What is the total amount payable over a 10-year period for
      i   Alison
      ii Graham?

   c   Alison and Graham look at the term life insurance table closely. They notice that the monthly premium for a woman is always lower than for a man of the same age. Why do you think this is?

Key point

An iterative formula is generated by a simple relation of the form
   $x_{n+1} = f(x_n)$
Each term in the sequence is calculated using the value of the previous term.

2 Instead of paying monthly insurance premiums, Alison and Graham consider putting the money in a savings account. They use an iterative formula to calculate the value of their savings each year.

$$I_{n+1} = 1.05(I_n + x)$$

where I_{n+1} is the value of the savings in year $n + 1$

I_n is the value of the savings in year n

x is the money put into the savings account each year.

$I_0 = 0$

a How much money would they put into the savings account each month in total?

b How much money would they put in each year?

c What is the value of

  i I_1

  ii I_2?

d Complete the table to work out the value of I_{10}.

Year	Total savings
0	£0
1	
2	

You could use a spreadsheet package to complete the table.

3 **a** If Alison and Graham both live for the next 10 years, would it be better if they had chosen to take out a life insurance policy or to invest their money? Explain your answer.

b If one of Alison and Graham dies in the next 10 years, would it be better if they had chosen to take out a life insurance policy or to invest their money? Explain your answer.

4 After reading all this information, Alison says that life insurance is really just about probabilities.

Do you agree with Alison? Explain your answer.

Explore

What other types of insurance are there? How do probabilities work in these types of insurance?

What do insurance companies do with all the money they receive?

What does an actuary do?

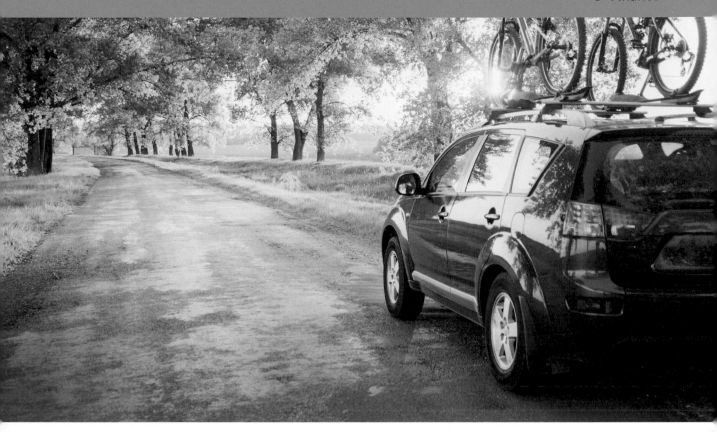

5.3 Car loans

Data source

Amir wants to buy a car. He needs £13 000 to help pay for it. He is exploring three alternative ways of paying for the car.

Credit card

Amir has a credit card that has a limit of £10 000. The APR of the card is 18.9%. The credit card company requires Amir to make a minimum payment of 4% of the balance each month. Amir can choose to pay more than the minimum balance if he wishes. He is charged interest on any balance that remains outstanding at the end of the month.

Loan

Amir has researched different loans. He has been pre-approved for a £13 000 loan with an APR of 4.3%. If Amir decides to take out the loan, he can choose from three different repayment period options, shown in Table 1.

Table 1 Options for repayment period on loan

	Option A	Option B	Option C
Repayment period	12 months	36 months	60 months
Monthly repayment amount	£1104.36	£383.73	£239.85

Technical literacy

The APR (annual percentage rate) is the annual rate of interest that is charged for borrowing. The APR is expressed as a percentage.

Technical literacy

The repayment period is the length of time over which you repay a loan.

Savings

Amir inherited some money in 2010. He immediately invested it in a fund that is linked to the stock market. Graph 1 shows how the fund performed between 2010 and 2015. Amir can withdraw money from this fund at any time without paying a fee.

Graph 1 Value of Amir's inheritance investment

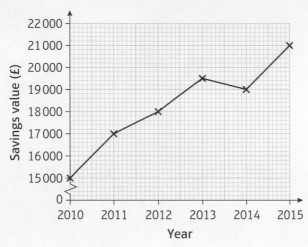

Look at the data source

1   Give two reasons why using the credit card to pay for the car would not be a good idea.

2   The formula to convert an annual percentage rate, A, into a monthly rate, M, is

$$M = (1 + A)^{\frac{1}{12}} - 1$$

where both rates are given as decimals.

  a   What is the annual rate for the credit card?
      Write your answer as a decimal.

  b   What is the monthly rate for the credit card?
      Write your answer as a decimal to 3 significant figures.

  c   What is the monthly rate for the loan?
      Write your answer as a decimal to 3 significant figures.

Key point

With compound interest, the interest is added to the initial investment, so that the added interest will also earn interest in the future. The formula for the value of an investment after n years is

$$\text{amount} = \text{initial amount} \times \left(\frac{100 + \text{interest rate}}{100} \right)^{n}$$

3 **a** How much money did Amir inherit?

  b For how many years had the inheritance been invested by 2015?

  c What was the value of the investment in 2015?

  d On average, what was Amir's annual interest rate?
    Give your answer to the nearest per cent.

  e Should Amir take out the loan or use money from his investment?
    Explain your answer.

Q3d hint

Substitute the values from Question 3a—c into the formula.

You can either use trial and improvement or rearrange the formula.

4 Amir decides to take out the loan.

  a How many years is each repayment period?

  b Calculate the total repayment amount for each option.

  c Explain why the total amount repayable is not just £13 000.

  d Work out how much money Amir will pay in interest for each option.

  e State one advantage and one disadvantage of paying off the loan in 12 months as opposed to 36 months.

5 Amir works out an iterative formula to calculate how much money will be outstanding on his loan every month if he chooses Option A:

$$R_{n+1} = 1.003\,515(R_n - 1104.36)$$

where R_n is the amount outstanding in month n

  R_{n+1} is the amount outstanding in month $n+1$

  R_0 is the amount at the beginning of the loan.

  a What is the value of R_0?

  b Work out the value of R_1 using the iterative formula.

  c Complete the table, showing the amount outstanding after each month.

Month	Amount outstanding
0	£13 000
1	
2	

  d Use the data from Question 5c to draw a graph for the amount outstanding.

Q5b hint

Each term in the sequence is calculated using the value of the previous term.

Q5cd hint

You could use a spreadsheet to complete the table and to draw the graph.

Explore

Investigate alternative ways of funding a payment of £10 000.

What factors affect the repayment amount? What different financing APRs can you find? Do they come with any incentives, e.g. no payment necessary for 12 months?

5.4 Mortgages

Data source A

Lisa is a 26-year-old website designer living in the South East of England. She has been commuting to her job as a website designer in London since leaving university 3 years ago. Currently, Lisa's annual salary is £27 000. Her goal is to buy a house before she is 30. Table 1 lists Lisa's current assets and liabilities.

Table 1 Lisa's assets and liabilities

	Asset	Liability
Current account	£1900	
Savings	£7400	
Student loan		£19 500
Car loan		£4100

After tax, national insurance and student loan repayments, Lisa takes home £1750 per month. Table 2 shows how much Lisa pays out regularly each month.

Table 2 Lisa's monthly outgoings

Item	Expense
Rent	£475
Car loan	£170
House bills	£150
Food	£275
Travel	£200
Social	£150

Lisa has started to look at houses in the local area and has found one that interests her. Its listed price is £208 000. She has contacted a few lenders about obtaining a mortgage and has found lenders that require a 5%, 10% or 20% deposit.

Technical literacy

Assets are items you own that have monetary value.

Liabilities are amounts of money that you owe someone else.

Once Lisa has put down the deposit, the mortgage lenders need to be confident she is capable of making monthly payments on the remaining balance. For this reason, the lenders will only allow Lisa to borrow up to 3.5 times her annual salary.

Data source B

Both house prices and the rate at which they change varies significantly across the country. The highest prices are in London and the South East, while prices in parts of the North are much lower. During the economic slowdown from about 2010, house prices grew very slowly or even fell in some areas. However, this changed in about 2013, with prices rising in all regions except the North East, as shown in Table 3.

Table 3 Average house prices and rate of change of house prices by region, September 2015

Region	Average price (£)	Annual change (%)
London	499 997	9.6
South East	256 737	8.5
East	211 161	8.3
West Midlands	141 470	4.8
South West	193 915	4.4
East Midlands	136 643	3.8
Yorkshire and Humberside	124 473	3.7
North West	115 594	2.5
Wales	119 916	0.7
North East	99 559	−0.3

Source: Land Registry Open Data

Data source C

Lisa has researched website designers' salaries in London. She found a survey of 60 website designers together with their annual salaries. These are displayed in a grouped frequency table in Table 4.

Table 4 Website designer salaries in London

Salary, x (£)	Frequency
$15\,000 < x \leqslant 25\,000$	13
$25\,000 < x \leqslant 35\,000$	15
$35\,000 < x \leqslant 45\,000$	11
$45\,000 < x \leqslant 55\,000$	8
$55\,000 < x \leqslant 65\,000$	8
$65\,000 < x \leqslant 75\,000$	5

Data source D

Lisa is considering two options for where to place her money while she saves for a house: a savings account or an investment fund linked to the stock market.

Savings account

Lisa is currently keeping her money in a savings account. The savings account has an AER of 1.2%. Lisa is worried that the outlook for interest rates does not look good, and she might be prepared to take a little more risk with her money in order to get a higher return.

Investment fund

Lisa has researched investment funds that are linked to the stock market. Table 5 shows the return (percentage increase in value of the investment) in each of the last 3 years from the fund that she prefers.

Table 5 Returns on an investment fund over the past 3 years

Year	Annual return
2012	9.7%
2013	4.5%
2014	4.9%

Lisa knows that in this fund her money could increase or decrease, so she has to decide whether she is prepared to accept this element of risk.

Technical literacy

The AER (Annual Equivalent Rate) is the annual rate of interest that savers receive on money in a particular account. The interest is paid not only on the original sum of money, but also on the interest already earned. The AER is expressed as a percentage.

Look at the data sources

Using Data source A

1. **a** What is the total of Lisa's monthly expenses? Do not include tax and student loan repayments.

   b What is the maximum amount she can pay into savings each month?

   c Lisa plans to buy a house in 4 years' time.
   How much money can she add to her savings by the end of 4 years?

   d What assumptions have you made when calculating your answer to part c?

   e Lisa already has some money in her savings account.
   How much money will she have saved altogether in 4 years' time?

   f Lisa says that the amount calculated in part e is less than she would expect to have in total. Why?

Using Data source B

2 a What was the annual percentage increase in house prices in Lisa's area in 2015?

  b Assume house prices in Lisa's area continue to grow at the annual rate shown in Table 3.
  Calculate the expected price of the house Lisa is interested in buying in 4 years' time.

  c Will Lisa be able to use her savings to put down a 5%, 10% or 20% deposit?

  d If Lisa decides to put down a 5% deposit, how much money will she need to borrow?

  e How much would Lisa have to earn in order to be able to borrow that amount of money?

  f Comment on the likelihood of Lisa being able to buy a house in 4 years in her present situation.

Using Data source C

3 a Draw a cumulative frequency graph for the data in Table 4.

  b Use the cumulative frequency graph to create a box plot for the salary data.

Lisa has had excellent performance reviews for the last few years.
She says she is confident she can move into the top 25% of salaries.

  c What salary could Lisa expect to earn in this scenario?

  d Lisa expects to take home about 72% of her new salary after tax, national insurance and student loan repayments. How much is this
  i per year
  ii per month?

  e If Lisa's other expenses remain the same, how much money would she be able to add to her savings
  i each month
  ii over 4 years?

4 Lisa's parents have offered for her to move in with them while she saves to buy a house. They would require her to pay bills (including food bills), but not any rent.

Assuming that she pays the same in bills as when she was renting, how much money could Lisa now add to her savings

  a each month

  b over 4 years?

Using Data source D

5 Lisa moves in with her parents and finds a new job that pays £53 000.

   a How much money in total can she save
     i per month
     ii per year?

   b Lisa decides to put 25% of her savings into a savings account and 75% into a stock market investment fund. How much money each year goes into
     i the savings account
     ii the investment fund?

   c The savings account grows annually using the following iterative formula:
$$V_{n+1} = 1.012(V_n + x)$$
     where $V_0 = 7400$
            x = annual investment.
     i Why is $V_0 = 7400$?
     ii Work out V_1, V_2, V_3 and V_4.

   d Lisa assumes the stock market fund will grow each year by the mean return for the previous 3 years.
     Work out the mean return. Round your answer to the nearest tenth.

   e Write an iterative formula for the value of Lisa's stock market fund investment, using the same format as for the savings investment.

   f Explain why $V_0 = 0$ in this case.

   g Work out V_1, V_2, V_3 and V_4.

6 At the end of 4 years, Lisa finds a house for £290 000.

   a Use your answers to Questions 5cii and 5g to work out how much Lisa has for a deposit.

   b Can Lisa put down a 20% deposit? Explain your answer.

   c How much money would Lisa need to borrow if she put down the 20% deposit?

   d Would she be allowed to borrow that amount on her new salary?

   e Banks can make exceptions when lending money if they think their client represents a good risk. Explain why Lisa represents a good risk.

7 It is often said that it is very difficult for young people in the UK to be able to buy a house.
After working through this project, do you agree or disagree with this statement?
Give evidence to support your answer.

Explore

Research the long-term relationship between wages and house prices. What does this tell you about the difficulties of trying to buy a first house?

What policies has the Government introduced in attempts to make it easier for first-time buyers?

6 Creative arts

In this unit you will work on projects to see how maths is used in the creative arts.

You will:

- Investigate Fibonacci-style sequences and the Golden Ratio
- Find out why certain combinations of musical notes sound good when played together
- Understand how sequences are used in music software.

Prior knowledge

Before beginning this unit, make sure you can:

- Plot algebraic graphs
- Write an expression for the nth term of a sequence
- Relate ratios to fractions

6 Creative arts

6.1 Ratios and art

Data source A

Paper is manufactured in various sizes although there is an international standard set out by the International Organisation for Standardisation (ISO). The series of paper sizes starts at a base A0, which has an area of $1\,m^2$. The following sizes in the series, A1, A2, A3 and so on, are found by halving the previous paper size across the larger dimension.

Data source B

Have you seen the sequence of numbers 1, 1, 2, 3, 5, 8… before? It is the start of the Fibonacci sequence, which was introduced to Europe by Leonardo of Pisa, an Italian mathematician born in the 12th century. He developed the sequence as an explanation for population growth of rabbits. The rule for the Fibonacci sequence is that each term is the sum of the previous two terms. Fibonacci-style sequences using the same rule can be generated using different starting numbers.

The Fibonacci sequence has been found to occur in many ways and forms in nature.

The Fibonacci sequence is also related to a specific ratio that is known as the Golden Ratio, or ϕ (the Greek letter phi), which was first recorded by the Ancient Greeks. For centuries, it has been claimed that artists and architects have been using the Golden Ratio to create works of art and buildings that look aesthetically pleasing.

Two quantities, a and b, where $a > b > 0$, are in the Golden Ratio (ϕ) if

$$\frac{a}{b} = \frac{a+b}{a}$$

A golden rectangle, often used in architecture, has its sides in this ratio.

Look at the data sources

Using Data source A

1 **a** Measure the sides of a sheet of A4 paper. Let a be the long side and b be the short side.

   b Is this paper size in the Golden Ratio? Explain how you know.

Using Data source B

2 **a** Copy and complete this table to find the first 15 terms of the Fibonacci sequence that starts 1, 1 ...

Term number	Term value	Ratio of term : previous term
1	1	—
2	1	$\frac{1}{1} = 1$
3	2	$\frac{2}{1} = 2$
4		

   b Plot a graph of the calculated ratio against the term number.

   c Explain what happens to the ratio as the term number increases.

3 Repeat Question 2 for another Fibonacci-style sequence, using any two positive integers as the first two numbers.
Plot your graph on the same axes as in Question 2.

4 **a** Repeat Question 2 for another Fibonacci-style sequence, using one positive integer and one negative integer as the first two numbers.

   b Explain what you notice.

   c Repeat parts a and b, using two negative integers as the first two numbers.

5 In the diagram, ABCDE is a regular pentagon. The five-pointed star that has been made by extending the sides is called a pentagram.

   a Work out these ratios, measuring to the nearest mm.

   i $\frac{JG}{JD}$ **ii** $\frac{JD}{JE}$ **iii** $\frac{JE}{ED}$

   b Comment on the values you found in part a.

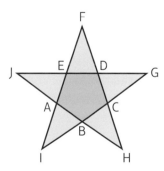

Explore

Use the internet to investigate the use of the Golden Ratio in

- The Last Supper by Leonardo da Vinci
- The Birth of Venus by Sandro Botticelli.

What did the Renaissance artists call the Golden Ratio?

6.2 Making music

Data source

Sound moves through the air in waves as air particles vibrating backwards and forwards in the same direction as the wave. This creates regions of high and low pressure that travel from the sound source to the listener. In music, the pitch or note you hear depends on the frequency of the waves. Frequency is measured in hertz (abbreviation Hz), which is the number of vibrations in each second. For example, the note A in the middle octave has a frequency of 440 hertz, or 440 waves per second.

In Western music, every note has a letter from A to G. A major or minor scale has eight notes, with the start and end notes an octave apart, for example, eight notes from C to C. A chromatic scale uses all the notes, including all the black keys on a piano.

A scale can start on any note, but the first scale that you usually learn on the piano is C major because it only uses the white notes, as shown in Figure 1. The chromatic scale starting on C has the notes C C# D D# E F F# G G# A A# B C.

Figure 1 Part of a piano keyboard

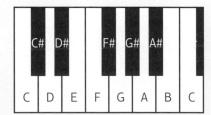

Every octave repeats these notes, but at a different pitch, so that C played in two different octaves is the same note but at a different pitch. The pitch of a note is determined by the frequency of the vibration. When one note

> ## Technical literacy
>
> Two notes are an octave apart if one has twice the frequency of the other.

is exactly an octave above another note, the frequency is doubled. The octave can be shown with a subscript from 0 to 8. The frequency of C_1 is twice that of C_0.

Table 1 shows the frequencies of the notes in the middle octave of a piano keyboard (usually denoted by a subscript 4) when A_4 is taken as 440 hertz.

Table 1 Frequencies of notes C_4 to B_4

Note	Frequency (Hz)
C_4	261.63
$C\#_4$	277.18
D_4	293.66
$D\#_4$	311.13
E_4	329.63
F_4	349.23
$F\#_4$	369.99
G_4	392.00
$G\#_4$	415.30
A_4	440.00
$A\#_4$	466.16
B_4	493.88

Musicians discovered a long time ago that some musical notes sound good played together, whereas others do not. Pairs of notes whose frequencies are in a ratio involving small numbers sound better than pairs of notes where the frequency ratio uses large numbers. For example, a frequency ratio of 4 : 3 sounds better than a ratio of 19 : 11.

Look at the data source

1   Using Table 1, calculate the frequency for each note in the third octave.

2   The time period for one vibration of the air particles (one wave) can be calculated from $T = \dfrac{1}{f}$, where T is the period in seconds and f is the frequency in hertz.

   a   Calculate the time period of the wave for the note C_4. Give your answer to 4 significant figures.

   b   The note C_4 is played on a guitar. Plot a graph showing when each of the first 24 vibrations occurs. Plot the number of the vibration on the x-axis and the elapsed time on the y-axis.

   c   Calculate the time period of the wave for the note C_3.

Q2b hint

Use crosses to plot the points so they still show after you have drawn the line joining them.

d On the same axes as for part b, plot the time of each of the first 15 vibrations of the string when the note C_3 is played.

e Give a reason why C_4, while higher, might have a similar sound to C_3.

f **i** Sketch on your axes the line that shows the time of each vibration for the note an octave lower, C_2.

  ii What is the equation of this line?

g Explain why each line must go through the origin.

h Is it possible for any line representing the arrival time of a vibration to be parallel to the line for C_4?
Explain your reasoning.

3 **a** Calculate the time period of the waves for the notes E_4 and G_4.

b Plot them on the axes from Question 2.

c Calculate the frequency ratios for the following pairs of notes. Express each one as an integer ratio in its simplest form.

  i $G_4 : C_4$

  ii $E_4 : C_4$

d Explain why the notes C, E and G sound good together.

Key point

The wavelength of a sound wave is the distance over which the wave repeats.

The wavelength of a sound wave can be calculated by using the formula $\lambda = \dfrac{v}{f}$, where λ (Greek lambda) is the wavelength in metres (m), f is the frequency in hertz (Hz) and v is the speed at which the wave is moving in metres per second (m/s).

4 **a** The speed of sound is approximately 345 m/s.
Copy and complete this table to show the wavelength of each note of the fourth octave in centimetres.

Note	Frequency (Hz)	Wavelength (cm)
C_4	261.63	131.87
$C\#_4$	277.18	
D_4	293.66	
$D\#_4$		

b Draw the graph of wavelength against frequency.

c What type of function does the graph represent?

5 **a** Work out the frequencies of the C notes from C_0 to C_8.

 b Draw the graph of frequency against octave number.

 c What type of function does the graph represent?

 Most humans can hear noises down to a frequency of about 20 hertz.

 d Which is the lowest C that most humans can hear?

 e Estimate the lowest note that most humans can hear.

6 The notes in a chromatic scale form a geometric sequence.

 a Taking the frequency of C_4 as the first term in the sequence, find the common ratio.

 b Calculate the 24th term in this sequence.

 c Humans can hear up to approximately 20 000 hertz. Using the sequence described in part a, would a human be likely to hear the 80th note in this scale?

Explore

In the scale of C major you have already seen that the notes C, E and G work well together – these notes form a chord.
Are there any other notes that would also sound good in this scale?

Q5b hint

Plot octave number on the x-axis and frequency on the y-axis.

Q5e hint

Look at your graph. On the x-axis, divide the distance between C notes by 12 so you can read off the notes in-between.

Q6a hint

For a geometric sequence,
$a_n = a_1 \times r^{n-1}$, where a_1 is the first term of the sequence and r is the common ratio.

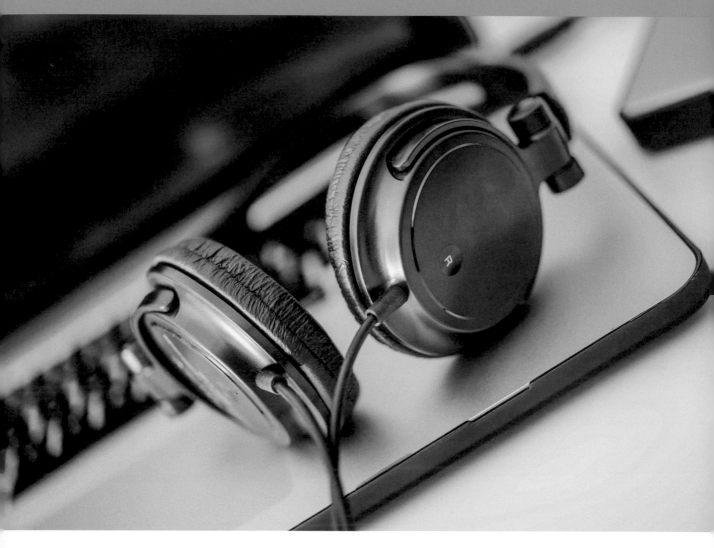

6.3 Music software

Data source

Nowadays, music composers have software to create the rhythm and beat in their music. The software enables them to use a series of mathematical sequences to create the complex rhythms within it.

Think about a piece of music that is made up from a series of numbered beats, starting at 1 and continuing to 320.

The software can be programmed to play a particular sound on every 4th beat, while another sound plays on every 6th beat and so on.

Josh wants to create a piece of music using an arithmetic sequence so that a bass drum sounds on the 1st beat, the 4th beat, the 7th beat and so on.

He adds a snare drum that sounds on the 3rd beat, the 7th beat, the 11th beat and so on.

He also uses a special effect which plays according to the sequence generated by $a_n = 5n + 2$.

Look at the data source

1 Write an expression for the nth term of Josh's initial composition, with only the bass drum.

2 Write an expression for the nth term of the sequence for the snare drum.

3 On which beats will the special effect be played?

4 All three sounds will be heard together on the 7th beat.
On which beat will they all be heard next?

5 Josh decides to use the snare drum only 20 times.
Calculate the beat on which the 20th snare drum will sound.

6 After 15 bass drum sounds Josh decides to change the pitch of the drum.
Calculate the beat on which the new bass drum sound will be heard.

7 Assume that Josh has made the changes described in Questions 5 and 6.
Which sounds can be heard on

  a the 167th beat

  b the 262nd beat?

Explore

The time signature of a piece of music tells you how the music is to be counted.

A time signature of $\frac{4}{4}$ means there are 4 'quarter notes' (crotchets) in each bar, so the beat is counted **1**, 2, 3, 4, **1**, 2, 3, 4 and so on. A time signature of $\frac{3}{4}$ means there are 3 quarter notes in each bar.

Explore the time signatures of some well-known pieces of music.
Do different styles tend to have different time signatures?

Q1 hint

In an arithmetic sequence,
$a_n = a_1 + (n-1)d$
where a_1 is the first term of the sequence and d is the common difference.

Q5 hint

Substitute $n = 20$ in the expression for the snare drum sequence.

Q7a hint

Substitute $a_n = 167$ into each expression and solve for n. Is n an integer?

7 Health

In this unit you will work on projects which explore contagious diseases and pain medication.

You will:

- Analyse the effect of vaccinations on the spread of disease
- Compare the incidence of disease in different parts of the world
- Investigate how the amount of medication in the body changes over time.

Prior knowledge

Before beginning this unit, make sure you can:

- Draw a probability tree diagram
- Calculate percentages
- Use averages to compare sets of data.

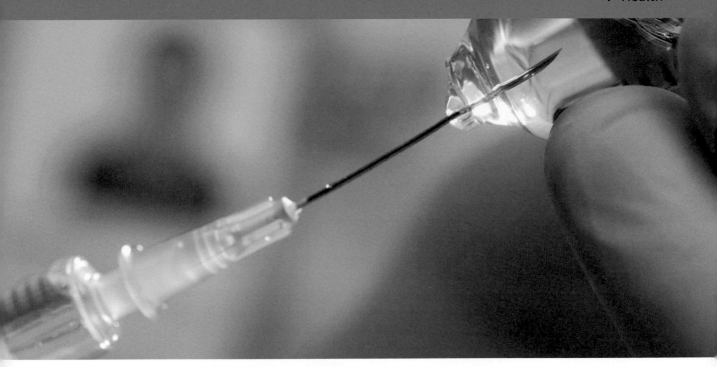

7.1 Measles and vaccination

Data source A

Measles is a highly infectious viral disease, which can lead to serious complications. In high-income regions of the world, such as Western Europe, it causes death in around 1 in 5000 cases, but as many as 1 in 100 cases of measles cause deaths in the poorest regions of the world.

Worldwide, measles is still a major cause of death, especially among children in resource-poor countries. However, over the last 20 years vaccination has dramatically reduced the number of deaths.

Before a vaccine existed, there were often hundreds of thousands of measles notifications every year. In 1967, the year before a vaccine was first introduced, there were 460 407 notified cases of measles in the UK, and 99 people died from the disease. In 1997, there were 3962 notified cases of measles in the UK and 3 deaths.

Data source B

If a child who is not immunised comes into contact with someone who has measles, it is very likely that they will catch the disease and risk developing serious complications. Out of a group of 100 people, about 90 will catch measles and 7 of those will have complications.

Vaccination is very effective, but it is not 100 per cent preventative and some vaccinated people are still at risk of contracting the disease. It is administered in two doses, the first at one year old and the second at 4 or 5 years old. The first dose provides 95 per cent protection and the second gives 99 per cent protection.

Table 1 shows the percentage of one-year-olds who were vaccinated against measles in the UK from 1980 to 1990.

Table 1 Percentage of one-year-olds vaccinated against measles in the UK, 1980–1990

Year	1980	1981	1982	1983	1984	1985	1986	1987	1988	1989	1990
%	53	55	58	60	63	68	71	76	80	84	87

Source: Based on data from the World Health Organization

Data source C

Table 2 shows the number of cases of measles in 45 European countries in 2014.

Table 2 Number of reported cases of measles in 45 European countries, 2014

Albania	0	Portugal	0	Serbia	37
Andorra	0	San Marino	0	Slovenia	52
Austria	0	Slovakia	0	Romania	59
Bulgaria	0	Ukraine	0	Belarus	64
Croatia	0	Vatican City	0	Belgium	70
Estonia	0	Greece	1	Macedonia	116
Finland	0	Republic of Moldova	2	United Kingdom	133
Hungary	0				
Iceland	0	Norway	3	Netherlands	140
Italy	0	Cyprus	10	Spain	154
Liechtenstein	0	Lithuania	11	Czech Republic	222
Luxembourg	0	Switzerland	23	France	267
Malta	0	Sweden	26	Germany	443
Monaco	0	Denmark	27	Bosnia and Herzegovina	3000
Montenegro	0	Ireland	33		
Poland	0	Latvia	36	Georgia	3188

Source: Based on data from the World Health Organization

Look at the data sources

Using Data source A

1 Find the percentage change in the number of notified cases of measles in the UK from 1967 to 1997. Interpret your answer in context.

2 In 1967 the population of the UK was estimated to be 54 959 000.
In 1997 the population of the UK was estimated to be 58 314 200.
   a Work out the percentage of the population with a notified case of measles in 1967.
   b Work out the percentage of the population with a notified case of measles in 1997.
   c Compare your answers to parts a and b.

Using Data source B

3 **a** Use a probability tree diagram to work out the probability that a one-year-old child would catch measles in 1980.

b Use a probability tree diagram to work out the probability that a one-year-old child would catch measles in 1990.

c Compare your findings from parts a and b.

4 **a** Use a probability tree diagram to work out the probability that a 10-year-old child would catch measles in 1990.

b Compare your answer from part a with Question 3b.

c What assumption did you make in part a?

5 **a** If 500 one-year-olds were exposed to measles, how many would you expect to catch the disease

   i in 1980

   ii in 1990?

b Of those infected, how many would you expect not to have been vaccinated

   i in 1980

   ii in 1990?

c What proportion of those infected would you expect not to have been vaccinated?

Q3a hint

There are two possibilities. A child could be vaccinated and catch measles or not vaccinated and catch measles. Use the effectiveness of the first dose of the vaccine.

Q4a hint

A 10-year-old child in 1990 would have been a one year old in 1981.

Using Data source C

6 This table shows some statistics for the numbers of cases of measles in 54 African countries in 2014.

Mean	1640.611
Median	58
Interquartile range	501
Standard deviation	5245.444

a Calculate the mean, median, interquartile range and standard deviation for the numbers of cases of measles in European countries in 2014.

b Compare the numbers of cases of measles for Europe and Africa.

Explore

Comment on your findings in relation to the information in the data sources about vaccination and catching measles.

What other factors need to be taken into account when investigating the connection between vaccination and the number of cases of measles?

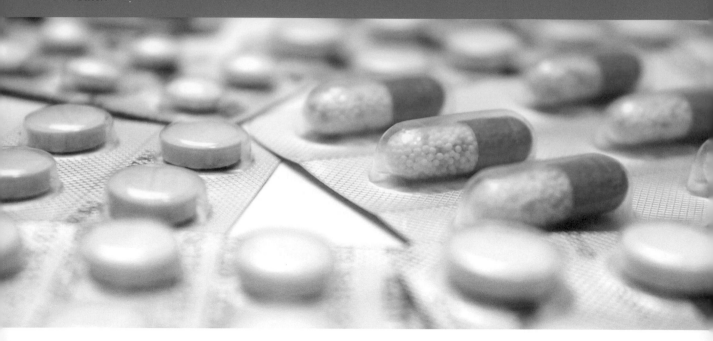

7.2 Paracetamol

Data source A

Paracetamol is a painkilling medicine that is available to buy over-the-counter and does not require a doctor's prescription. It has been widely available since the 1950s and is used to treat headaches and high temperature.

One tablet usually contains 500 milligrams of paracetamol. The recommended dose for an adult or child over 12 is one to two tablets every four hours. The maximum dose in a 24-hour period is eight tablets.

When taken at the recommended dosage, paracetamol is usually safe and effective. However, taking more than this can be harmful.

Paracetamol is quickly absorbed into the digestive system. Graph 1 shows how the amount of paracetamol in the body after a single dose of two tablets might decrease over time.

The decrease in amount of paracetamol in the body can be modelled using this formula:

$$C = 1000 \times 2.2^{-0.352t}$$

where C is the number of milligrams of paracetamol in the body and t is the number of hours since the dose was taken.

Graph 1 Variation of amount of paracetamol in body over time after a single dose

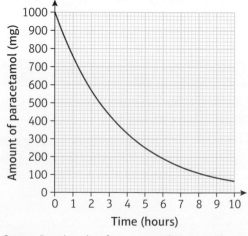

Source: Based on data from www.medicines.org.uk

Data source B

If a person takes further doses of paracetamol after 4, 8 and 12 hours, you could expect the level of paracetamol in their body to be as shown in Graph 2.

Graph 2 Variation of amount of paracetamol in body over time after repeated doses

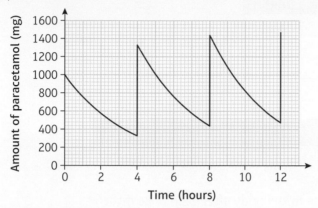

This iterative formula can be used to work out how much paracetamol will be in the body when the person takes another 1000-milligram dose of paracetamol every 4 hours:

$$C_{n+1} = 0.33\,C_n + 1000$$

where $C_0 = 1000$ and n = the number of 4-hour intervals.

Look at the data sources

Using Data source A

1   Use Graph 1 to find the half-life of paracetamol in the body.

2   A person takes one dose of 1000 milligrams of paracetamol.
    a How many milligrams of paracetamol would you expect to be in their body after 5 hours?
    b Approximately how long will it take before the amount of paracetamol in the person's body is less than 10 milligrams?

3   Use the formula to answer these questions.
    a Find the amount of paracetamol in a person's body 6 hours after a single dose of 1000 milligrams.
    b How long will it take for the amount of paracetamol in the person's body to fall to 10 milligrams?
    c Compare your answer to part b with the value you found in Question 2b.

Technical literacy

The half-life is the time taken for the amount of paracetamol in the body to decrease by half.

Q2b hint

You could use a spreadsheet to work out the level after every half-life.

Using Data source B

4   What proportion of a paracetamol dose will be left in the body after 4 hours?

5   Use the iterative formula to work out the amount of paracetamol in the body immediately after the person takes the next 3 doses. Copy and complete this table.

n	Number of hours	C_n (mg)
0	0	1000
1	4	
2		
3		

Explore

What would happen if the person took the paracetamol more frequently, for example every 2.5 hours?

What amount of paracetamol would be in the body after each dose if the person continued to take 1000 milligrams of paracetamol every 4 hours for 24 hours?

8 Economy

In this unit you will work on projects involving payday loans, goods trading and product sales.

You will:

- Examine the costs involved with payday loans
- Identify trends in UK imports and exports
- Analyse changes in vinyl record sales.

Prior knowledge

Before beginning this unit make sure you can:

- Use formulae
- Calculate probabilities
- Calculate averages.

8 Economy

8.1 Payday loans

Data source A

Payday loan companies provide borrowing options to people who need to borrow money for a short period only. The payday lender will pay the requested amount directly into the payee's bank account. The payee then has a specified amount of time to pay the loan back with interest.

Payday loans usually have particularly high interest rates compared with other borrowing options such as credit cards (which are typically 17% to 30% APR) or a bank loan (typically 3% to 30% APR). Payday lenders often charge in excess of 1000% APR.

Graph 1 shows the amount owed on a payday loan of £500 over 12 months from a company which sets its APR at 1500%.

Graph 1 Amount owed on a payday loan of £500

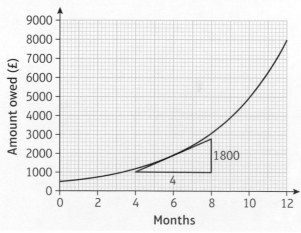

Technical literacy

The lender is the loan company. The payee is the person borrowing the money.

Technical literacy

The annual percentage rate (APR) is the annual interest rate that is charged for borrowing.

Graph 1 can be used to estimate the rate at which the amount owed increases per month by working out the gradient of the tangent to the curve at any point.

For example, at month 6

$$\text{rate of increase} = \frac{1800}{4} = £450 \text{ per month}$$

The following formula can be used to convert from an annual percentage rate, R%, to a monthly rate, r%:

$$\frac{r}{100} = \left(1 + \frac{R}{100}\right)^{\frac{1}{12}} - 1$$

Data source B

In July 2014 the Financial Conduct Authority (FCA) published its proposals for a cost cap on payday loans. These are the terms:

- Initial cost cap of 0.8% per day. This lowers the cost for most borrowers. For all high-cost short-term credit loans, interest and fees must not exceed 0.8% per day of the amount borrowed.
- Fixed default fees capped at £15. This protects borrowers struggling to repay. If borrowers do not repay their loans on time, default charges must not exceed £15. Interest on unpaid balances and default charges must not exceed the initial rate.
- Total cost cap of 100%. This protects borrowers from escalating debts. Borrowers must never have to pay back more in fees and interest than the amount borrowed.

From 2 January 2015, no borrower will ever have to pay back more than twice what they borrowed, and someone taking out a loan for 30 days and repaying on time will not pay more than £24 in fees and interest per £100 borrowed.

Source: Adapted from Financial Conduct Authority

Data source C

Credit unions are financial cooperatives and offer an alternative way to borrow money. They are non-profit organisations set up by members with something in common, for example working in the same industry or living in the same area. They are run by the members for the benefit of the members.

A credit union interest rate is dependent on the applicant's credit history, the amount borrowed and the repayment plan. The APR on a typical credit union loan is from 12% to a maximum of 42.6%. Although interest rates on personal loans from high-street banks are usually lower than this, bank loans are only available to applicants with an excellent credit rating.

Technical literacy

A cooperative is an organisation that is owned and run by its members, who share the profits.

Data source D

A research organisation carried out a survey of 400 people who had used some form of credit. Table 1 shows the numbers of full-time workers (FT), part-time workers (PT) and students (S) who had taken out a payday loan (PL) or some other form of credit (OC).

Table 1 Participants in a credit survey

	Payday loan (PL)	Other credit (OC)
Full-time workers (FT)	35	115
Part-time workers (PT)	40	110
Students (S)	25	75

Look at the data sources

Using Data source A

Key point

A tangent to a curve at a point is a straight line that just touches the curve at that point:

It can be used to work out the gradient of a curve at a particular point.

1   Sean takes out a payday loan of £400 at a rate of 24% per month.

   a He makes no repayments. Work out how much he owes after each of the first 12 months.

   b Work out the percentage increase in the amount owed after 12 months.

   c Draw a graph to show how much Sean owes over 12 months.

   d Work out the rate of increase in the amount owed after
      i 4 months   **ii** 8 months.
      Compare your answers.

2   Use the formula to work out the monthly rate for a payday loan with an APR of 1500%.

3   Use the formula to work out the APR for a monthly rate of 24%. Compare your answer with your answer to Question 1b.

Using Data source B

4   A payday loan website has this offer.

Borrow £350!

Make 52 weekly repayments of £13.39

Work out the APR. Comment on your answer.

Q1di hint

Draw the tangent to the curve at 4 months.

Extend the ends of the tangent to points that will give an integer value for the base of the triangle.

5 Use the rules introduced by the Financial Conduct Authority (FCA).
Work out the maximum amount that can be paid back to the lender in
Question 4 for:

a a £500 loan for 30 days that is repaid on time

b a £500 loan for 15 days that is repaid on time

c a £500 loan that the payee fails to pay back after 100 days.

Using Data source C

6 Kate takes out a credit union loan of £400 at an APR of 12.7%.
She doesn't make any repayments in the first year.
Work out how much she owes at the end of the first year.

7 Work out the monthly rate of interest for a loan with an APR of 12.7%.

8 For the loan in Question 6

a draw a graph to show the amount owed over 12 months

b work out the rate of increase in the amount owed after
i 4 months
ii 8 months.

9 Compare Sean's loan in Question 1 with Kate's loan in Question 6.

Using Data source D

Key point

Set notation is used to show combinations of outcomes.
For outcomes A and B:
- $P(A \cup B)$ means the probability of A or B or both
- $P(A \cap B)$ means the probability of A and B
- $P(A')$ means the probability of not A
- $P(B|A)$ means the probability of B given that A is true.

10 A person is selected at random from the research organisation's survey
sample. Work out

a P(PT)

b P(FT ∩ PL)

c P([FT ∩ PL]′)

d P(PL|S)

Q10a hint

P(PT) is the
probability that this
person is a part-time
worker.

Explore

Work out the smallest number of days it would take for the amount owed
on a payday loan of £100 to become double the loan amount. Assume
that the FCA's rules are followed.

8.2 Imports and exports

The economy in the UK is very reliant on global trade.

When the UK buys goods from another country, these are imports. The UK's biggest imports are transport (cars), minerals (oil), medical drugs and electronics (computers), and it also imports food, clothing, precious metals and much more.

When the UK sells goods to another country, these are exports. The UK's biggest exports are transport (cars), minerals (oil), medical drugs and machinery (gas turbines), but the UK has many other exports, including chemicals, metals and precious metals.

Tables 1 and 2 give the values of UK imports from and exports to the European Union between 2007 and 2014.

Table 1 UK imports from the European Union, 2007–2014

Year	2007	2008	2009	2010	2011	2012	2013	2014
Imports from EU (£ millions)	167 821	178 858	161 634	184 726	201 599	206 914	217 130	218 503

Source: HM Revenue and Customs UK trade info

Table 2 UK exports to the European Union, 2007–2014

Year	2007	2008	2009	2010	2011	2012	2013	2014
Exports to EU (£ millions)	126 920	141 068	124 649	141 931	158 293	149 986	150 425	146 783

Source: HM Revenue and Customs UK trade info

Data source B

Figure 1 shows the proportions of the value of total UK imports and exports, of the imports to and the exports from each separate country of the UK in the first three months of 2014.

Figure 1 Share of UK imports and exports by UK country, January to March 2014

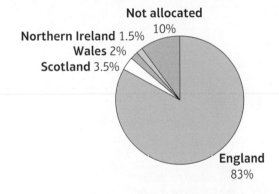

Share of UK's imported goods

Not allocated 10%
Northern Ireland 1.5%
Wales 2%
Scotland 3.5%
England 83%

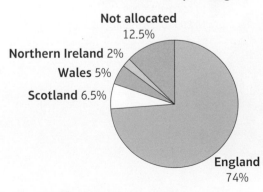

Share of UK's exported goods

Not allocated 12.5%
Northern Ireland 2%
Wales 5%
Scotland 6.5%
England 74%

Source: Based on data from the Office for National Statistics

Data source C

Governments find it useful to look at productivity data at different levels of detail. For example, the UK government can analyse data at the level of the whole UK, by country, by region, by county or by smaller districts. In order to compare how productive different countries, regions or districts are, measures are divided by the relevant population.

Table 3 shows the distribution of gross value added (GVA) per capita according to district in December 2013. Figure 2 shows GVA per capita by region on a map of the UK.

Table 3 UK GVA per capita by district, 2013

GVA per capita (£ thousands)	Frequency (number of districts)
$11 < x \leq 17$	50
$17 < x \leq 18$	10
$18 < x \leq 20$	28
$20 < x \leq 22$	10
$22 < x \leq 26$	20
$26 < x \leq 41$	21

Source: Based on data from the Office for National Statistics

Technical literacy

Gross Value Added (GVA) is a measure of the cost of producing goods and services less the costs directly associated with the production. GVA per capita is the GVA divided by the population.

Figure 2 UK GVA per capita by region, 2013

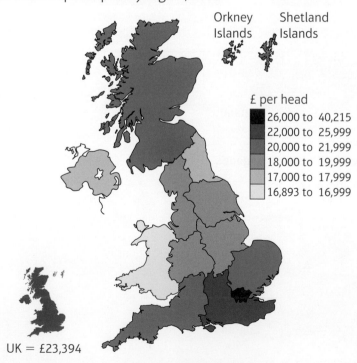

Source: Based on information from the Office for National Statistics

Look at the data sources

Using Data source A

1 a Plot a time series graph to show the changes in the value of UK imports and exports from 2007 to 2014. Plot both lines on the same axes.

b In which 12-month period did the gap between UK imports and exports begin to widen?

2 a Use your graph to describe the UK's import and export trade over this period.

b What is a possible explanation for the dip in trade in 2009?

Using Data source B

3 Which is the only UK country to have a higher proportion of imports than exports in the first quarter of 2014?

4 Assume that the value of goods is proportional to the number of goods. What is the probability, in the first quarter of 2014, of

a a randomly selected, exported good originating in Wales

b a randomly selected, imported good going to Scotland?

5 Sharon looks at Table 2 and Figure 1 and says, 'In 2014 England made £108 619.42 from exporting goods to the European Union'. What values has she used and what assumptions has she made?

Using Data source C

Key point

You can draw a histogram for data with unequal class intervals by plotting 'frequency density' instead of 'frequency' on the y-axis. Using frequency density ensures that the height of each bar isn't unreasonably large compared with the actual frequency, regardless of the class interval.

$$\text{Frequency density} = \frac{\text{frequency}}{\text{class width}}$$

6 Construct a histogram for the data on GVA per capita from Table 3.

7 You now have two visual representations showing GVA per capita in 2013. How could they be interpreted differently?

8 Explain why an estimate of the mean may be a better average to use than an estimate of the median for the data in Table 3.

Explore

Question 4 asked you to assume that the value of goods is proportional to the number of goods. Is this a valid assumption?

What factors contribute to the growth of the UK economy?

Explore hint

Find out the main import and export items for each geographical area and the value of these items.

8.3 Vinyl record sales

Data source

Developments in technology mean that there are now many ways to buy and access music. In the past, music has been sold in many formats, including vinyl records, compact cassettes and compact discs. More recently, digital downloading and music streaming have dominated the global market. Table 1 shows how the number of vinyl records sold in the US changed from 1972 through to the beginning of the 21st century.

Table 1 US sales of vinyl records, 1972 to 2000

Year	Vinyl record sales (millions of units)	Year	Vinyl record sales (millions of units)
1972	510	1987	140
1973	480	1988	60
1974	420	1989	35
1975	465	1990	20
1976	525	1991	15
1977	520	1992	12
1978	500	1993	11
1979	470	1994	10
1980	450	1995	13
1981	370	1996	9
1982	330	1997	8
1983	335	1998	7
1984	290	1999	6
1985	210	2000	5
1986	180		

Source: Based on estimates from the Recording Industry Association of America (RIAA)

Look at the data source

1 **a** Describe the trend in sales between 1972 and 2000.

   b Was the fall in sales of vinyl records linear? Explain.

2 Juan suggests the following model for vinyl sales from 1976 to 2000:

$$S = 525 \times 0.80^{x - 1976}$$

where S = sales in millions of units

   x = year.

   a Interpret the values 525 and 0.80 in the context of this scenario.

   b Use the model to estimate sales of vinyl records in 1980, 1985, 1990, 1995 and 2000.

   c Comment on the accuracy of Juan's model.

3 Emily suggests that sales have fallen by only about 12% each year.

   a Write a model for a 12% annual drop in sales.

   b Test this model for sales in 1980, 1985, 1990, 1995 and 2000.

   c Whose model do you think more accurately represents vinyl sales? Explain.

Key point 1

The sum of the first n terms of a geometric series is

$$S_n = \frac{a(1 - r^n)}{1 - r}$$

where a = first term

   r = common ratio

   n = number of terms.

Key point 2

If the common ratio of a geometric series is less than 1, then each term is smaller than the previous term. However many terms of the series you add together, the sum will never exceed a certain number. This number is called the sum to infinity of the series. The symbol for infinity is ∞.

The sum to infinity of a geometric series is

$$S_\infty = \frac{a}{1 - r} \qquad |r| < 1$$

where a = first term

   r = common ratio.

4 Assume that the data in Table 1 can be modelled by a geometric series.

   a Use Juan's model to calculate the total sales in the period

   i 1976 to 1980

   ii 1976 to 1990.

   b Use Emily's model to calculate the total sales from 1976 onwards.

Explore

Investigate how vinyl sales have changed since 2000.

9 Travel

In this unit you will work on projects involving car stopping distances, international travel and tourism.

You will:

- Investigate the links between speed and thinking, braking and stopping distances
- Explore aspects of travelling abroad
- Examine trends in the amount spent by overseas visitors to the UK and UK residents travelling abroad.

Prior knowledge

Before beginning this unit make sure you can:

- Use formulae
- Work with scatter diagrams
- Calculate averages.

9.1 Stopping distances

Data source

Table 1 shows the typical thinking and braking distances for a car travelling at various speeds.

Table 1 Thinking and braking distances at speeds from 20 to 70 miles per hour

Speed (miles per hour)	Thinking distance (metres)	Braking distance (metres)
20	6	6
30	9	14
40	12	24
50	15	38
60	18	55
70	21	75

Source: Adapted from 'Typical Stopping Distances' from www.gov.uk

Technical literacy

The thinking distance is the distance the car travels between the driver seeing an obstacle and applying the brakes.

The braking distance is the distance the car takes to stop once the driver applies the brakes.

Look at the data source

1 **a** Explain how you can use a formula to find the total stopping distance from the thinking distance and the braking distance.

   b Calculate the total stopping distance for a car travelling at each of the speeds in Table 1.

   c Draw a grid with speed on the x-axis and distance on the y-axis. On the same grid, plot a line or curve for
     i thinking distance
     ii braking distance
     iii total stopping distance.

   d Is the relationship between speed and thinking distance linear? Explain.

e Is the relationship between speed and braking distance linear? Explain.

f You are going to investigate how the stopping distance changes as the speed changes.

  i Draw a tangent to the stopping distance graph at 35 mph and calculate the gradient of the tangent.

  ii Draw a tangent to the graph at 55 mph and calculate the gradient of the tangent.

g Comment on the effect of speed on stopping distances and the change an increase of 1 mph can make.

2 In order to be able to calculate stopping distances at different speeds more easily, you could create a formula to model the situation.

  a Mark suggests using a linear model.
  Explain why a linear model would probably not be appropriate for modelling stopping distances.

  b Jane suggests using a quadratic model with the following formula:

  $$y = 0.0157x^2 + 0.2629x + 0.6$$

  where x = speed in miles per hour and y = stopping distance in metres.

  Use Jane's model to calculate the stopping distance for each of the speeds in Table 1.

  c Dwight suggests using a different model with the following formula:

  $$y = 0.0826x^{\frac{5}{3}}$$

  where x = speed in miles per hour and y = stopping distance in metres.

  Use Dwight's model to calculate the stopping distance for each of the speeds in Table 1.

  d Which model do you think best fits the data? Explain.

3 Speed in miles per hour can be converted into speed in metres per second using this formula:

$$\text{speed in metres per second} = \frac{\text{speed in miles per hour} \times 4}{9}$$

Use the formula to convert each of the speeds into metres per second.

4 A road safety campaign aims to encourage drivers to think about, and hopefully to reduce, their speed.

They want to calculate how fast a car would hit a ball if it rolled straight out into the road 15 m in front of the moving car. They create this formula:

$$v = \sqrt{u^2 - 13.2(15 - \tfrac{2}{3}u)}$$

where u = initial speed in metres per second and v = speed after 15 m in metres per second.

a Explain why the formula is not required for an initial speed of 20 mph.

b Explain why the formula is not appropriate for speeds above 50 mph.

c Calculate the speed of impact for initial speeds of 25, 30, 35, 40 and 45 mph

 i in metres per second

 ii in miles per hour.

d Create a new formula for a ball rolling out at a distance of 10 m in front of the car, by replacing 15 with 10 in the formula above.

e Calculate the speed of impact using the formula from part d for initial speeds of 20, 25, 30, 35, 40 and 45 mph

 i in metres per second

 ii in miles per hour.

f Create a new formula for a ball rolling out at a distance of 20 m.

g Calculate the speed of impact using the formula from part f for initial speeds of 20, 25, 30, 35, 40 and 45 mph

 i in metres per second

 ii in miles per hour.

h Parts of some towns now have 20 mph speed limits instead of 30 mph speed limits. Use the values you have calculated to explain the importance of lower speed limits.

5 a On the same axes, plot the impact speed in miles per hour against the initial speed for a ball rolling out at distances of 10 m and 20 m in front of a moving car.

b Dawn says the difference in impact speeds is much greater at lower initial speeds.

 Explain how the graph supports her statement.

6 On motorways, large arrows (called chevrons) are sometimes marked to encourage drivers to keep a safe distance apart.

The chevrons are spaced at a distance of 40 m, and drivers are encouraged to keep at least two chevrons (80 m) apart.

Comment on whether this is a safe distance.

Explore

How do different road conditions affect stopping distances?

9.2 International travel

Max lives in Reading and regularly travels to New York on business. Today, Max's flight leaves at 4.15 pm, and he must be at the departure gate at Heathrow Airport at least 60 minutes before the plane's scheduled departure time.

Max walks from his flat to Reading train station, which takes him between 8 and 10 minutes. Once at the train station he can take any of the trains to London Paddington station shown in Table 1.

Table 1 Times of trains from Reading to London Paddington

Departure time	Journey time
13 09	34 minutes
13 17	37 minutes
13 25	32 minutes
13 36	41 minutes
13 41	36 minutes

Once Max arrives in London Paddington, it takes him between 5 and 6 minutes to walk from his train from Reading to the Heathrow Express train. Trains depart at 10, 25, 40 and 55 minutes past the hour. The journey to Heathrow takes 15 minutes.

From the train, Max walks to the airport terminal, a walk of between 7 and 10 minutes. As a frequent flyer, he can join an express lane to drop off his luggage. This takes between 5 and 10 minutes. Then he joins the express security lane, which takes between 10 and 15 minutes. From the security lane, Max walks to the departure gate, a walk of between 1 and 8 minutes.

Data source B

Max flies direct from London Heathrow to John F. Kennedy International Airport in New York. His flight time on the way there (London–New York) is 7 hours and 55 minutes. His flight time on the return journey is 6 hours and 40 minutes.

The reason the times are different is the jet stream, a current of fast-moving air at high altitudes. Although the plane speed is exactly the same, the outward flight is into a headwind. This means that the journey time is longer. On the return trip the plane is flying with the wind and the journey is quicker.

The formula for the speed of the journey from London to New York is

speed from London to New York $= p - w$

where $p =$ speed of plane and $w =$ wind speed.

The distance from London to New York is 3465 miles or 5576 kilometres.

Data source C

Max earns air miles for every flight he takes. He can use these air miles to take a future journey for free or at a reduced cost. Max would like to take a holiday in a big city. Table 2 shows some of the cities he is interested in visiting.

Table 2 Distances, flight times and ticket prices from London to different cities

Destination city	Distance (km)	Ticket price (£)
Cape Town	9682	1600
Buenos Aires	11 140	1920
Beijing	8 150	1550
Paris	344	90
Moscow	2503	825
San Francisco	8626	1560
Istanbul	2502	340
Sao Paulo	9508	1550
Mexico City	8939	1450
Toronto	5718	1170
Tokyo	9576	1240
Dubai	5480	1315

Look at the data sources

Using Data source A

1 **a** Draw an inequality on a number line for each part of Max's journey from leaving the Heathrow Express train to reaching the departure gate. Work backwards from the departure gate.

  i Walk to departure gate

  ii Time in security lane

  iii Time at luggage drop

  iv Walk from train to airport

  b How long will this part of the journey take Max

  i in the best-case scenario

  ii in the worst-case scenario?

  c Max likes to plan for the worst-case scenario.
  Explain why his train from Reading needs to arrive in London Paddington by 2.04 pm.

  d Which train should Max take from Reading to London Paddington?

  e At what time should Max leave his flat?

> ### Technical literacy
>
> A best-case scenario is the best result that can be expected.

Using Data source B

2 **a** Convert each journey time into a decimal number of hours.

  i From London to New York

  ii From New York to London

  b Explain why the formula for the speed of the journey from London to New York subtracts the speed of the wind.

  c Write down a formula for the speed of the journey from New York to London.

  d Write down a pair of equations relating speed, distance and time for the two journeys, assuming the wind speed is the same on both journeys.

  e Solve the equations to find the value of p and the value of w.

> ### Q2d hint
>
> distance =
> speed $\times$ time

Using Data source C

3 **a** Draw a scatter diagram with distance on the x-axis and the cost of the flight on the y-axis.

  b Describe the correlation between flight distance and cost.

  The equation of the regression line is $y = 0.145x + 224.62$.

  c Explain how the cost of a flight alters with each additional mile travelled.

  d The distance from London to Dallas, USA, is 7590 km.
  Use this model to estimate the cost of a flight to Dallas.

Explore

Are flights cheaper on certain days of the week?

Investigate budget airlines and their flights to Europe. Is there still a correlation between distance and cost?

9.3 Tourism

Tables 1 and 2 show information about the numbers of visits, nights visited and the total spending by overseas visitors to the UK and by UK residents travelling abroad between 2000 and 2013.

Table 1 Overseas visitors to the UK, 2000–2013

Year	Number of visits (thousands)	Number of nights (millions)	Total spending (£ millions)
2000	25 209	203.8	12 805
2001	22 835	189.5	11 306
2002	24 180	199.3	11 737
2003	24 715	203.4	11 855
2004	27 755	227.4	13 047
2005	29 970	249.2	14 248
2006	32 713	273.4	16 002
2007	32 778	251.5	15 960
2008	31 888	245.8	16 323
2009	29 889	229.4	16 592
2010	29 803	227.8	16 899
2011	30 798	235.2	17 998
2012	31 084	230.2	18 640
2013	32 692	245.5	21 258

Source: Data from Travel Trends (various editions), Office for National Statistics

Table 2 UK residents travelling abroad, 2000–2013

Year	Number of visits (thousands)	Number of nights (millions)	Total spending (£ millions)
2000	56 837	566.9	24 251
2001	58 281	578.8	25 332
2002	59 377	595.2	26 962
2003	61 424	618.1	28 550
2004	64 194	645.9	30 285
2005	66 441	669.0	32 154
2006	69 536	701.3	34 411
2007	69 450	689.6	35 013
2008	69 011	703.3	36 838
2009	58 614	614.5	31 694
2010	55 562	607.0	31 820
2011	56 836	594.7	31 701
2012	56 538	584.2	32 450
2013	57 792	604.9	34 510

Source: Data from Travel Trends (various editions), Office for National Statistics

Data source B

Table 3 shows the average spending per night by visitors to the UK between 1980 and 2009. The data is adjusted for inflation.

Table 3 Average spend (adjusted for inflation) made by visitors to the UK, 1980s, 1990s and 2000s

1980s	Average spend per night (£)	1990s	Average spend per night (£)	2000s	Average spend per night (£)
1980	38.98	1990	62.30	2000	81.30
1981	41.34	1991	61.26	2001	75.66
1982	43.21	1992	64.20	2002	73.23
1983	50.01	1993	74.37	2003	71.04
1984	53.03	1994	74.32	2004	68.57
1985	56.73	1995	76.25	2005	66.99
1986	59.91	1996	78.31	2006	67.23
1987	58.79	1997	75.53	2007	71.46
1988	58.68	1998	73.90	2008	73.33
1989	59.90	1999	77.88	2009	78.29

Source: Adapted from data from Travel Trends (various editions), Office for National Statistics

Technical literacy

Annual inflation is the yearly increase in the cost of goods or services. As inflation continues, the same amount of money buys less each year.

Data source C

The Channel Tunnel opened in 1994. For UK residents travelling overseas between 2000 and 2013, Table 4 compares the number travelling by ferry with the number using the Channel Tunnel.

Table 4 Visits made overseas by UK residents by ferry and using the Channel Tunnel, 2000–2013

Year	Ferry (thousands)	Channel Tunnel (thousands)
2000	9646	5799
2001	9651	5619
2002	10 038	5349
2003	9200	5123
2004	8950	4809
2005	8102	4713
2006	8411	4665
2007	8473	4649
2008	8145	4825
2009	7559	4398
2010	8056	4267
2011	7857	4255
2012	6755	4867
2013	6922	4845

Source: Data from Travel Trends (various editions), Office for National Statistics

Look at the data sources

Using Data source A

1   **a** Overall, who spent more money: people visiting the UK or UK residents travelling abroad?

   b Explain how you can use the data to calculate average spending per night.

   c Work out the average spending per night of overseas visitors coming to the UK for each year from 2000 to 2013.

   d Calculate the three-point moving averages for average spending per night of visitors coming to the UK.

   e Plot the values from parts c and d on a graph.

   f Comment on the shape of the graph.

Q1d hint

First three-point moving average

$$= \frac{2000 \text{ spend} + 2001 \text{ spend} + 2002 \text{ spend}}{3}$$

Plot the average centred on 2001.

Key point

To make the data more comparable, the spending can be adjusted for inflation. We can assume that inflation is about 2% per year on average. To adjust a figure so that it can be compared with 2013 data, use the formula

spending adjusted for inflation = (actual spending) $\times 1.02^{\text{(years before 2013)}}$

2 a Explain why it might be unfair to compare spending in 2000 with spending in 2013 directly.

b Calculate the spending in 2012 adjusted for inflation for UK visitors travelling abroad (Table 2).

c Complete the table to show spending adjusted to 2013 prices.

Q2b hint

2012:
£32 450 $\times 1.02^1 = \square$

Year	Number of visits (thousands)	Number of nights (millions)	Total spending (£ millions)	Total spending adjusted for inflation (£ millions)	Spend per night adjusted for inflation (£ millions)
2000	56 837	566.9	24 251		
2001	58 281	578.8	25 332		

d Calculate the adjusted spending per night for each year.

e Interpret your results from part d.

Using Data source B

3 a Calculate the minimum, lower quartile, median, upper quartile, maximum and interquartile range for each decade.

b Draw three box plots to represent this data, one for each decade. Draw the box plots one above the other, using the same scale. This will make comparisons easier.

c Comment on the box plots.

d Do the data sets contain outliers?

e Calculate the mean and the standard deviation for each decade and comment on your results.

Q3d hint

lower limit for outliers
= LQ – 1.5 × IQR

upper limit for outliers
= UQ + 1.5 × IQR

Using Data source C

4 **a** Draw time series graphs for the numbers of journeys by ferry and using the Channel Tunnel.
   b Draw on two trend lines (lines of best fit), one for each data set.

The product moment correlation coefficients (PMCCs) for the data sets are:

PMCC for year and number of ferry journeys $= -0.938$

PMCC for year and number of Channel Tunnel journeys $= -0.757$

   c Interpret the correlation between
   i year and number of ferry journeys
   ii year and number of Channel Tunnel journeys.

The equations of the least squares regression lines are

Ferry: $V = -223.4x + 456\,739$

Channel Tunnel: $V = -84.1x + 173\,548$

where $V =$ number of journeys (in thousands) and $x =$ year.

   d Interpret the gradient of these regression lines.
   e According to the formulae, in what year will the numbers of people travelling by ferry and through the Channel Tunnel be equal?
   f Comment on the possible sources of error in these calculations.

Explore

Investigate the number of UK residents travelling to different countries outside Europe.

Is any country seeing a decline?

Q4a hint

Plot both data series on the same grid.

Q4c hint

The PMCC measures the degree of correlation between two variables:
$1 =$ perfect positive correlation,
$-1 =$ perfect negative correlation,
$0 =$ no correlation.

Q4d hint

A regression line is the line that matches the pattern of data as closely as possible. The least squares regression line is the one that has the smallest possible value for the sum of the squares of the y-distances of all the data points from the line.

10 Environment

In this unit you will work on projects involving climate, deforestation and green technology.

You will:

- Investigate deforestation
- Consider the cost of environmentally friendly technology
- Analyse climate and weather data.

Prior knowledge

Before beginning this unit, make sure you can:

- Find the equation of a line through two given points
- Find the rate of change at a point on a curve
- Plot a scatter diagram.

10.1 Deforestation

Data source

Every year, the Amazon rainforest in Brazil is reducing in size due to deforestation. The causes of this deforestation are a combination of clearance for mining raw materials, clearance for wood stocks and clearance for farming and ranching.

Graph 1 Total area of forest cover in Brazil, 1990–2010

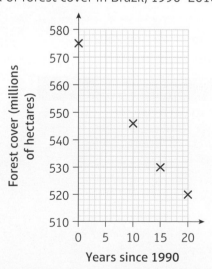

Technical literacy

Deforestation is the clearing of an area of trees, usually by cutting down or burning.

Source: Based on Brazil forest information and data from Mongabay.com

Look at the data source

1 a What was the total area of forest cover in Brazil in 1990?

  b What was the total area of forest cover in Brazil in 2000?

  c Find the equation of the straight line through these two points.

  d Draw the straight line on a copy of Graph 1.
    Is the line a good representation of the rate of deforestation in the Brazilian rainforest?
    Give a reason for your answer.

  e Repeat parts c and d for
    i the first and third points
    ii the first and fourth points.

  f Give an interpretation of the gradients of these lines.

2 An ecology team claims that the rate of deforestation is declining and that the actual model should be quadratic.
They propose a relationship between x and y such that

$$y = 0.0245x^2 - 3.279x + 575.2$$

where y is the total area of forest cover and x is the number of years after 1990.

  a Work out the values for total forest cover in each of 1990, 2000, 2005 and 2010 using this model.
    Give your answers to the nearest million.

  b Using the same scales as in Graph 1, draw a graph of the quadratic relationship proposed by the ecology team.
    Comment on the suitability of the model.

  c Using this model, what would the total area of forest cover be in 2020?

  d What would the total area of forest cover be in 2100?
    Comment on your answer.

  e Draw a tangent to the curve at the point representing 2002.
    What was the rate of change in total forest cover in 2002?

Explore

Explore what measures have been put in place to decrease the global rate of deforestation.

10.2 The cost of going green

Data source

The term 'carbon neutral' is used to describe a situation where there is no net release of carbon dioxide into the atmosphere. It can be brought about by carbon-offsetting, where businesses reduce carbon dioxide output in some areas to compensate for emissions elsewhere.

Domestically, households can reduce their 'carbon footprint' by switching to energy-saving alternatives. These will reduce their demand for energy supplied from sources with high levels of carbon emissions, such as traditional coal- or oil-fired power stations.

Table 1 shows the costs and approximate annual savings of several green energy solutions that could be introduced into the home.

Table 1 Energy-saving alternatives: costs and savings

Measure	Cost (£)	Saving per year (£)
Solar water-heating panels	3000–5000	60–100
New 'A' rated boiler	1000–1500	100–120
Double glazing	2000–3000	70–100
Cavity wall insulation	250–300	100–150
Loft insulation	250–350	100–150
Draught proofing	80–100	30–40

An incentive scheme encourages households to fit solar electricity panels. The scheme involves a grant of 20% of the cost of the panels together with a low-interest loan to pay for the rest. The terms for the loan are a compound interest rate of 2% per year over a loan period of 10 years. There is also a use the term 'feed-in tariff' instead of 5p per kWh (kilowatt hour) for each kWh of electricity not used by the household.

Technical literacy

A household's carbon footprint is the amount of greenhouse gases, specifically carbon dioxide, emitted as a result of that household's activities during a given period.

Look at the data source

1   Jo and Steve recently bought an old house with single-glazed windows, an old boiler and no insulation.
They want to investigate how they can save money on energy bills while 'going green'.

   a   Using the information in Table 1, work out which measure will pay for itself the quickest. Explain your answer.

   b   Which measure will take the longest to pay for itself? Explain your answer.

   c   What other factors might affect their decisions?

2   Jo and Steve get a quote of £4500 for solar electricity panels.

   a   How much will they get as a grant from the incentive scheme?

   b   If they pay for the rest using the low-interest loan and pay nothing back during the period of the loan, how much will they owe at the end?

   c   They decide to pay £200 per year.
   How much will they owe at the end of
   i   the first year
   ii   the third year
   iii   the tenth year?

   d   Draw a graph of the amount they owe against the number of years of the loan. What type of graph is it?

   e   Assuming they pay nothing during the period of the loan, how many kWh of electricity must they sell back to the scheme to meet the cost of the panels?

   f   How many kWh must they sell back to meet the cost of the panels if they pay £200 per year?

   g   Steve calculates that his annual saving per year from having the panels installed is £80.
   After how many years will the panels pay for themselves? Explain how you know.

3   Jo says that getting solar panels is a waste of money and they would be better off just insulating the house and getting a new boiler.

   a   After how many years will the savings from these improvements cover the costs?

   To pay for the new boiler, Steve and Jo would need to take out a bank loan for the whole amount at a compound interest rate of 8% for 5 years.

   b   What is the maximum price they could pay for a boiler if the maximum monthly repayment they could afford was
   i   £20
   ii   £25?

Q2b hint

For a compound annual interest rate of i%:
final amount after n years = amount of loan $\times (1 + \frac{i}{100})^n$

Q2c hint

Interest is calculated *after* the repayment is made.
The amount owed at the end of the year is given by the iterative formula
$L_n = (L_{n-1} - 200) \times 1.02$

Q3b hint

Interest is calculated at the end of the year.

Key point

The nth term of a quadratic sequence has the form $an^2 + bn + c$.

To work out the nth term:

* work out the first and second differences between the terms
* halve the second difference to work out the coefficient of n^2 (a)
* subtract an^2 from the terms of the sequence (you may need to add a constant, or find the nth term of the remaining sequence)
* create a table to keep track.

Example

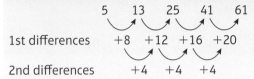

$$5 \quad 13 \quad 25 \quad 41 \quad 61$$

1st differences $\quad +8 \quad +12 \quad +16 \quad +20$

2nd differences $\quad +4 \quad +4 \quad +4$

$a = 4 \div 2 = 2$

So the sequence is $2n^2 + bn + c$.

n	1	2	3	4	5
Sequence	5	13	25	41	61
$2n^2$	2	8	18	32	50
Sequence – $2n^2$	3	5	7	9	11

The nth term of 3, 5, 7, 9, 11 is $2n + 1$.

The nth term of the quadratic sequence is $2n^2 + 2n + 1$.

4 A local double-glazing company offers a deal where customers pay off the cost of their new windows according to this formula:

monthly payment (£) = number of months after installation $\times$ 10 + 20

 a Jo and Steve decide to install double glazing costing £2500.
  i Write down their monthly payments for the first six months.
  ii Describe the type of sequence these payments make.
  iii How much will they have paid back in total during the first six months?
  iv Work out how many months it will take them to pay back the total cost.

 b Jo tells the company that they can pay back £10 at the end of the first month, £30 at the end of the second month, £60 at the end of the third month, £100 at the end of the fourth month, etc.
  i What type of sequence is this?
  ii Work out a formula for the nth term for these repayments, where a_n is the repayment at the end of the nth month.
  iii How long will it take Steve and Jo to pay off the windows using this formula?

10.3 Climate and weather

Data source A

Brisbane is a city in Queensland, Australia. Its climate is referred to as sub-tropical. The weather in Brisbane is typically hot sunny days combined with tropical storms in summer and a temperate, drier winter.

Tables 1 and 2 show the monthly rainfall and monthly average temperature for a city with a similar climate to Brisbane from 2000 to 2014.

Technical literacy

A temperate climate has temperatures that are not extreme, so are neither too hot nor too cold.

Table 1 Monthly rainfall in mm, 2000–2014

Year	Jan	Feb	Mar	Apr	May	Jun	Jul	Aug	Sep	Oct	Nov	Dec
2000	38.7	65.2	87.7	39.5	56.6	43.1	35.3	12.3	1.2	43.9	92.3	143.0
2001	26.6	202.2	247.8	33.5	36.1	21.2	28.6	8.0	23.9	98.3	191.5	147.4
2002	38.5	58.5	103.8	48.2	54.3	62.4	0.3	108.6	11.0	44.9	48.5	125.6
2003	7.1	133.7	80.4	113.3	81.8	50.1	21.0	20.1	19.5	49.9	28.3	143.8
2004	281.4	132.0	102.7	29.1	20.9	14.7	0.8	19.5	21.6	59.0	161.9	219.8
2005	89.2	34.6	14.5	46.3	53.9	90.3	11.2	23.8	15.2	128.0	133.1	78.4
2006	176.4	106.5	76.7	46.6	12.2	43.7	32.5	56.0	55.7	23.1	83.1	81.4
2007	97.2	55.5	18.3	5.6	39.8	111.9	3.2	93.1	34.5	55.5	59.6	80.4
2008	184.3	217.6	39.4	17.9	51.2	121.5	91.6	16.5	54.0	55.1	328.8	61.7
2009	78.1	132.6	47.5	196.4	238.3	88.9	3.0	2.3	21.6	57.0	32.7	173.3
2010	40.4	270.3	162.1	36.7	61.6	7.1	26.9	104.9	101.8	306.7	57.4	480.3
2011	315.5	158.4	151.3	98.5	68.3	8.3	12.7	74.2	16.8	115.8	13.2	132.6
2012	345.2	161.2	116.0	149.4	36.5	122.8	50.3	2.8	5.3	20.5	117.9	50.9
2013	281.5	250.3	173.1	103.2	44.6	58.8	37.0	0.4	15.1	22.3	104.6	18.5
2014	144.7	17.4	166.8	12.4	30.8	18.7	16.3	93.6	27.6	5.8	144.3	122.7

Table 2 Monthly average temperature in °C, 2000–2014

Year	Jan	Feb	Mar	Apr	May	Jun	Jul	Aug	Sep	Oct	Nov	Dec	Annual
2000	27.6	28.7	29.2	25.2	23.9	23.2	22.8	23.5	28.5	25.3	26.0	29.6	25.2
2001	31.0	28.9	29.5	27.6	25.0	21.9	24.3	22.6	23.0	26.1	29.8	31.0	26.8
2002	31.1	29.6	30.6	27.4	24.6	24.7	22.1	21.7	25.2	26.5	28.8	28.2	28.7
2003	30.1	28.1	29.2	26.7	23.9	20.0	20.3	23.1	29.6	28.0	30.0	27.7	26.2
2004	31.0	32.9	29.5	29.3	26.5	22.1	25.1	23.4	25.2	29.3	29.5	27.0	27.4
2005	30.2	30.0	29.5	25.9	23.5	20.3	23.9	25.3	26.5	25.9	29.6	30.3	27.5
2006	28.6	29.4	29.4	27.8	24.0	21.8	20.6	24.7	24.2	24.9	26.5	26.1	26.6
2007	30.5	28.3	33.1	27.7	25.3	21.9	21.9	21.1	25.8	28.6	28.3	30.3	24.2
2008	31.2	28.1	27.8	26.5	22.3	24.0	18.6	19.2	26.1	22.8	28.5	32.5	25.2
2009	27.8	27.6	30.1	29.3	25.6	20.4	23.3	25.3	27.4	28.5	28.7	29.6	25.3
2010	30.8	30.4	30.0	25.3	22.1	20.5	22.4	19.6	23.5	25.3	28.8	27.8	25.0
2011	28.5	31.8	29.7	28.0	24.8	22.3	22.5	21.3	27.0	24.5	30.9	25.7	27.9
2012	29.6	31.3	30.0	24.4	22.5	20.0	21.8	25.0	25.4	29.3	26.0	28.1	26.8
2013	30.5	26.3	28.8	26.8	23.1	23.1	22.5	23.8	27.4	27.0	29.0	28.1	25.9
2014	30.5	30.0	29.8	30.5	23.3	25.2	22.4	20.4	26.3	27.0	31.8	30.7	27.5

Data source B

When people talk about 'global warming' they are referring to a long-term change in annual average temperature, compared with the average over previous years. Table 3 shows how the annual average temperatures in the same city as the one depicted by Data source A compares with the average temperature recorded between 1961 and 1990 (23.2 °C).

Table 3 Average annual temperature difference in °C, 1910–2010, using the average temperature of 23.2 °C

Year	Difference from 1961–1990	Year	Difference from 1961–1990
1910	−0.47	1965	0.07
1915	0.72	1970	−0.06
1920	−0.41	1975	−0.11
1925	−0.90	1980	0.60
1930	−0.61	1985	0.05
1935	−0.18	1990	0.29
1940	−0.29	1995	0.43
1945	−0.15	2000	−0.17
1950	−0.93	2005	1.28
1955	−0.35	2010	−0.04
1960	−0.59		

Data source C

The daily global solar exposure is the total amount of solar radiation that reaches the Earth's surface in a day (measured midnight to midnight) on a horizontal surface. Table 4 shows the temperature and daily average solar exposure for the city depicted by Data source A and Data source B in 2008.

Table 4 Air temperature and daily average solar exposure, 2008

	Jan	Feb	Mar	Apr	May	Jun	Jul	Aug	Sep	Oct	Nov	Dec
Mean maximum temperature (°C)	27.3	29.2	27.5	25.8	24.8	22.5	21.2	21.8	24.3	25.0	27.5	30.4
Mean daily global solar exposure (megajoules per m²)	21.0	20.5	19.4	16.8	13.4	11.2	11.6	16.3	18.3	23.2	21.0	24.7

Look at the data sources

Using Data source A

1 **a** Work out the mean monthly rainfall for the city over this period for the months of January and June.

  b Plot, on the same set of axes, time series graphs for monthly rainfall in 2006 and in 2014.

  c **i** Describe how the monthly rainfall changes in 2006.

    ii Describe how the monthly rainfall changes in 2014.

    iii Compare the changes in monthly rainfall for each year.

2 **a** **i** Plot, on the same set of axes, time series graphs for monthly average temperature in 2000 and in 2014.

    ii Compare the changes in average temperature for each year.

  b **i** Use the annual average temperature column in Table 2 to work out the median annual average temperature.

    ii Work out the lower and upper quartiles.

    iii Draw a box plot of annual average temperature.

    iv The box plot below shows the same information for Stratford-upon-Avon in England.

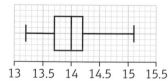

13 13.5 14 14.5 15 15.5

Average annual temperature (°C)

    Compare the annual average temperatures in the city depicted in the data sources and Stratford-upon-Avon.

3 **a** Work out the mean annual average temperature for the city depicted in the data sources.

b Work out the variance and standard deviation.

c Is it better to describe the monthly average temperature in this city using the median and interquartile range (worked out in Question 2), or the mean and standard deviation?

Q3 hint

Variance $= \frac{\sum(x - \bar{x})^2}{n}$

Standard deviation $= \sqrt{\text{variance}}$

Using Data source B

4 **a** What was the mean annual temperature in the city depicted in the data sources in

  i 1910

  ii 1980?

b Draw a time series graph to show the difference in average temperature from 1961–1990 between 1910 and 2010.

5 **a** Work out the four-point moving averages for the difference in temperature from 1961–1990 between 1910 and 2010.

b Plot the moving averages on the time series graph drawn in Question 4b.

c Draw a line of best fit through the moving averages.

d Describe the long-term trend in annual average temperature in the city depicted by the data.

Q5a hint

First four-point moving average

$= \frac{-0.47 + 0.72 + -0.41 + -0.9}{4}$

Plot the average centred on 1917.5

6 **a** **i** Relabel the x-axis in the time series graph, using 0 for 1910 up to 21 for 2010.

  ii Work out the gradient of your line of best fit.

  iii Work out the equation of your line of best fit.

  iv Use your line of best fit to predict the annual average temperature difference from 1961–1990 in 2020.

b Is there sufficient evidence to suggest that the Earth is suffering from 'global warming'? Why?

Using Data source C

7 Draw a scatter diagram for the data in Table 4.

8 **a** Work out the annual mean maximum temperature and annual mean daily solar exposure.

b Plot this point on the scatter diagram.
Make sure you clearly indicate it in a different way to the 12 data points.

c Draw a line of best fit through the mean point.

d **i** Predict the solar exposure when the temperature is 21.3 °C.

  ii Predict the solar exposure when the temperature is 35 °C.

  iii Which prediction is likely to be more accurate?
  Give a reason for your answer.

9 **a** Work out the equation of the least squares regression line.

   b Plot the regression line on your scatter diagram.

   c How close is the regression line to the line of best fit you drew in Question 8c?

   d Repeat Question 8di using your regression line. Compare the two values.

10 **a** Work out the product moment correlation coefficient (PMCC).

   b Interpret the value of this correlation coefficient. Discuss any limitations there might be from drawing conclusions based on the value of the correlation coefficient.

Q9a hint

The equation of the least squares regression line is given by $y = a + bx$

where $b = \dfrac{n\Sigma xy - \Sigma x \Sigma y}{n\Sigma x^2 - (\Sigma x)^2}$

$a = \bar{y} - b\bar{x}$

Q10a hint

$$\text{PMCC} = \dfrac{n\Sigma xy - \Sigma x \Sigma y}{\sqrt{n\Sigma x^2 - (\Sigma x)^2}\sqrt{n\Sigma y^2 - (\Sigma y)^2}}$$

Explore

Find similar weather data for real cities around the world. Do these statistics support the conclusions you made in this project?

11 Disasters

In this unit you will work on projects involving natural and man-made disasters.

You will:

- Interpret hurricane data
- Compare earthquake magnitudes
- Investigate the impact of fires.

Prior knowledge

Before beginning this unit, make sure you can:

- Substitute values into formulae
- Draw a scatter diagram.

11 Disasters

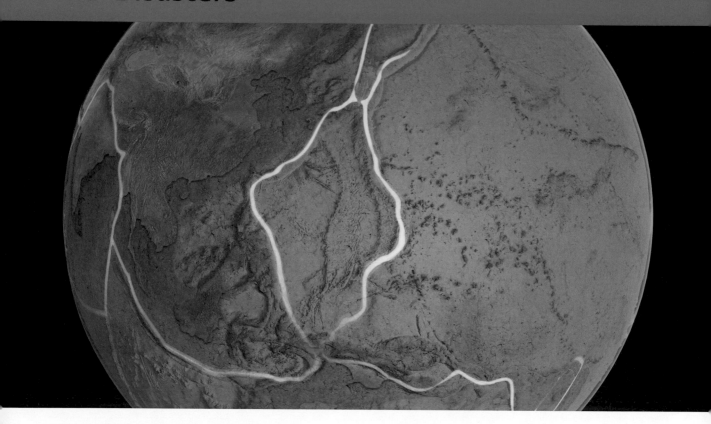

11.1 Earthquakes

Data source A

The magnitude, or size, of an earthquake is most commonly measured on the Richter scale. The Richter scale runs from 2 (an earthquake that is rarely felt) to 9 (an earthquake that causes total destruction). This scale is not linear: an earthquake that measures 6 on the Richter scale is not twice as strong as one that measures 3. In fact, the Richter scale is a base-10 logarithmic scale. This means that an increase of 1 on the Richter scale is actually an increase of 10 in the magnitude of the earthquake.

The relative strengths of two earthquakes can be compared using the formula

$$\text{relative strength} = 10^{M_2 - M_1}$$

where M_1 is the magnitude of the weaker earthquake and M_2 is the magnitude of the stronger earthquake.

Earthquakes release energy: this is what causes the destruction. The energy released by the earthquake, in joules, can be calculated using the formula

$$\text{energy} = 10^{4.8 + 1.5M}$$

where M is the magnitude of the earthquake.

Throughout history, there have been some big earthquakes that have caused a lot of damage and destruction. Table 1 lists the most powerful earthquakes experienced by ten countries this century.

Table 1 Largest earthquake in ten countries, 2000 to 2015

Country	Place	Date	Magnitude
Chile	Illapel	16 September 2015	8.3
China	Sichuan	12 May 2008	7.9
Greece	Kamariotissa	24 May 2014	6.9
Indonesia	Sumatra	26 December 2004	9.1
Italy	L'Aquila	6 April 2009	6.3
Japan	Tōhoku	11 March 2011	9.0
Mexico	Colima	22 January 2003	7.6
Pakistan	Awaran District	24 September 2013	7.7
UK	Market Rasen	27 February 2008	5.2
USA	Denali, Alaska	3 November 2002	7.9

Source: Based on data from USGS Earthquakes Hazard Program

Data source B

It is not yet possible to predict when earthquakes will occur. However, there is a pattern that links the size of an earthquake, as measured on the Richter scale, and the number of such earthquakes that occur each year.

Table 2 shows the total number of earthquakes of each magnitude that occurred in the ten-year period between 2005 and 2014.

Table 2 Number of earthquakes of each magnitude, 2005–2014

Magnitude range	Frequency
8.0–9.9	15
7.0–7.9	144
6.0–6.9	1525
5.0–5.9	17 768
4.0–4.9	117 126

Source: Based on data from USGS Earthquakes Hazard Program

Look at the data sources

Using Data source A

1   Use the respective magnitudes to compare the relative strengths of the UK and Indonesian earthquakes.

2   **a**   Calculate the energy released by

    i   the UK earthquake

    ii   the Indonesian earthquake.

   b   How many times as much energy was released by the Indonesian earthquake than by the UK earthquake?

3 There was some confusion over the strength of the Indonesian earthquake at first. At one point it was thought that it had a magnitude of 8.8

    a Compare the initial magnitude of this earthquake with the true reading, and comment on the relative strengths and the energy released.

    b A news reporter wants to explain to the public the difference between a magnitude 8.8 earthquake and a magnitude 9.1 earthquake. His initial idea is to say that there is not much difference. Explain how the news reporter should comment on this story.

Using Data source B

4 a Describe the pattern in the numbers of earthquakes of each magnitude. Is it a linear sequence?

    b Copy and complete the table, showing the logarithm of the total frequency of earthquakes with magnitude greater than or equal to M.

Q4b hint

Use the $\boxed{\log}$ button on your calculator.

Magnitude range	Frequency	Minimum magnitude, M	Frequency, f, with magnitude greater than or equal to M	$\log f$ (2 d.p.)
8.0–9.9	15	8.0	15	$\log 15 = 1.18$
7.0–7.9	144	7.0	159	

    c Plot a graph of $\log f$ against M.

5 a A regression line can be drawn on the graph. The regression line will have an equation of the form

$$\log f = a + bM$$

Find the equation of the regression line using $\Sigma M = 30$, $\Sigma \log f = 16.028$, $\Sigma M^2 = 190$, $\Sigma M \log f = 86.163$ and $n = 5$.

    b The equation of the regression line can be rearranged to

$$f = 10^{a + bM}$$

Rewrite your equation in this form.

6 a Use the regression line to predict the number of earthquakes of magnitude 3 or higher that would be expected to occur in a 10-year period.

    b Explain why using this formula with magnitude 3 earthquakes might not be valid.

Q5 hint

The equation of the least squares regression line is usually given by

$$y = a + bx$$

where

$$b = \frac{n\Sigma xy - \Sigma x \Sigma y}{n\Sigma x^2 - (\Sigma x)^2}$$

$$a = \bar{y} - b\bar{x}$$

Which variables have x and y been replaced by for your data?

Explore

Sound is often measured in decibels. How does this scale work? What are the similarities and differences between it and the Richter scale?

11.2 Hurricanes

Data source A

Table 1 lists some of the major (category 3 or higher) hurricanes in the Atlantic Ocean region between 2000 and 2012. Hurricanes are categorised according to their maximum sustained wind speed. Category 3 hurricanes have a wind speed of 178–208 km/h, category 4 have a wind speed of 209–251 km/h and category 5 have a wind speed of 252 km/h or higher.

The hurricane's name, maximum wind speed (in kilometres per hour) and minimum air pressure (in hectopascals) are also listed.

Technical literacy

Pressure is measured in pascals. The prefix hecto means 100, so 1 hectopascal (hPa) = 100 pascals (Pa).

Table 1 Major hurricanes in the Atlantic Ocean region, 2000–2012

Name	Year	Category of hurricane	Max wind speed (km/h)	Min pressure (hPa)	Damage ($ billions)
Michelle	2001	4	220	933	2.15
Isidore	2002	3	205	934	1.30
Isabel	2003	5	270	915	5.37
Ivan	2004	5	270	910	23.30
Charley	2004	4	240	941	15.10
Frances	2004	4	230	935	9.85
Jeanne	2004	3	195	350	7.66
Katrina	2005	5	280	902	125.00
Wilma	2005	5	295	882	29.30
Rita	2005	5	285	895	12.00
Dennis	2005	4	240	930	4.00
Dean	2007	5	280	905	1.78
Ike	2008	4	230	935	37.50
Gustav	2008	4	250	941	6.61
Karl	2010	3	205	956	5.60
Irene	2011	3	195	942	16.60
Sandy	2012	3	185	940	71.40

Source: Based on data from National Oceanic and Atmosphere Administration

Data source B

Table 2 shows how many named storms occurred in the Atlantic Ocean region each year since 2000. Tropical storms (one level down from a hurricane) are given a name when they display a rotating circular pattern and wind speeds of 39 mph (or 62 km/h). Once the wind speeds reach 119 km/h, the tropical storm is re-categorised as a hurricane. If the wind speeds subsequently reach 178 km/h, the hurricane becomes a major hurricane (category 3, 4 or 5, as described in Data source A). The Accumulated Cyclone Energy (ACE) index is a measure of the total intensity of all hurricanes during a particular year. The larger the ACE value, the higher the overall intensity of hurricanes that year.

Table 2 Number of named storms, hurricanes and major hurricanes in the Atlantic Ocean region, 1990–2014

Year	Named storms	Hurricanes	Major hurricanes	ACE
1990	14	8	1	97
1991	8	4	2	36
1992	7	4	1	76
1993	8	4	1	39
1994	7	3	0	32
1995	19	11	5	228
1996	13	9	6	166
1997	8	3	1	41
1998	14	10	3	182
1999	12	8	5	177
2000	15	8	3	119
2001	15	9	4	110
2002	12	4	2	67
2003	16	7	3	176
2004	15	9	6	227
2005	28	15	7	250
2006	10	5	2	79
2007	15	6	2	74
2008	16	8	5	146
2009	9	3	2	53
2010	19	12	5	165
2011	19	7	4	126
2012	19	10	2	129
2013	14	2	0	36
2014	8	6	2	67

Source: Adapted from National Oceanic and Atmospheric Administration, Hurricane Research Division

Look at the data sources

Using Data source A

1 a Explain why it is not appropriate to compare the cost of damages due to a hurricane in 2001 with the cost of damages due to a hurricane in 2012.

b Calculate the cost of damages from each hurricane in 2015 prices. Assume inflation is 2% per annum.

Q1b hint

cost in 2015 $=$ original cost $\times 1.02^{n}$ where $n =$ number of years before 2015

2 a Calculate the product moment correlation coefficient (PMCC) between maximum wind speed (x) and damages (y). Use the following values:

$\Sigma x = 4075$, $\Sigma y = 441.12$, $\Sigma x^2 = 997\,075$, $\Sigma y^2 = 34\,240.85$, $\Sigma xy = 109\,528.60$ and $n = 17$.

b Interpret your correlation coefficient. Comment on the relationship between the maximum wind speed and the cost of damages.

Q2a hint

$$PMCC = \frac{n\Sigma xy - \Sigma x \Sigma y}{\sqrt{n\Sigma x^2 - (\Sigma x)^2}\sqrt{n\Sigma y^2 - (\Sigma y)^2}}$$

3 a Draw a scatter diagram showing wind speed and damages.

b Use the scatter diagram to identify two outliers in the data.

c How would the correlation be affected if those outliers were removed?

Using Data source B

4 a Draw a time series graph showing the number of named storms each year between 1990 and 2014.

b Calculate the five-point moving averages.

c Plot the moving averages on the same graph.

d Do you think there is a trend in the number of named storms per year? Explain.

5 a Calculate the PMCC between the number of named storms (x) and the ACE (y). Use the following values:

$\Sigma x = 340$, $\Sigma x^2 = 5220$, $\Sigma y = 2898$, $\Sigma y^2 = 441\,824$, $\Sigma xy = 45\,397$ and $n = 25$.

b Comment on your correlation coefficient and the relationship between the number of named storms and the ACE.

c Is it reasonable to assume a link between the number of named storms and the ACE? Explain.

6 a Draw a scatter diagram with number of named storms on the x-axis and ACE on the y-axis. Calculate the equation of the regression line.

b In 2015, there were 12 named storms. Calculate the expected ACE.

Explore

Select ten of the hurricanes from Table 1. Find the biggest city each of them passed through. Is there any correlation between the population of that city and the cost of damages caused?

11.3 Fires

Data source

Table 1 shows the number and types of fires in the Greater London area for each year since 1980. Primary fires are fires that harm people or cause damage to property. Primary fires can occur indoors or outdoors. Secondary fires do not harm people or property and are usually outdoor fires, for example in grassland, but they may also occur in derelict buildings. Chimney fires are fires that are contained within the chimney of a building.

Table 1 Fires in the Greater London area, 1980–2014

Year	Primary fires	Secondary fires	Chimney fires	Total fires	Injury rate per million population
1980	19 571	26 493	581	46 645	139.8
1981	19 790	24 003	538	44 331	159.7
1982	20 551	24 162	502	45 215	161.0
1983	20 869	29 196	484	50 549	197.4
1984	21 133	29 504	439	51 076	179.1
1985	22 202	27 580	544	50 326	224.8
1986	22 119	23 521	430	46 070	229.8
1987	21 963	20 886	419	43 268	219.0
1988	22 550	24 789	394	47 733	223.1

continued

Year	Primary fires	Secondary fires	Chimney fires	Total fires	Injury rate per million population
1989	22 199	34 433	261	56 893	240.1
1990	21 635	34 155	204	55 994	223.7
1991	21 050	22 877	257	44 184	213.3
1992	20 684	20 732	222	41 638	218.0
1993	20 025	26 303	215	46 543	201.3
1994	19 080	28 463	150	47 693	219.8
1995	19 892	35 932	138	55 962	170.3
1996	20 414	31 380	165	51 959	231.0
1997	20 148	27 406	124	47 678	244.9
1998	19 677	21 295	99	41 071	248.1
1999	20 411	25 947	97	46 455	230.8
2000	22 334	26 135	85	48 554	189.2
2001	22 655	32 322	86	55 063	189.7
2002	20 271	28 213	60	48 544	182.3
2003	20 081	38 084	68	58 233	174.3
2004	17 788	23 023	72	40 883	158.2
2005	16 167	24 218	56	40 441	170.9
2006	15 373	21 674	66	37 113	196.7
2007	14 115	18 920	49	33 084	189.0
2008	13 372	16 211	70	29 653	161.5
2009	14 178	15 379	34	29 591	154.7
2010	13 522	13 895	50	27 467	152.8
2011	12 911	13 880	56	26 847	149.3
2012	11 678	9697	68	21 443	138.8
2013	11 289	9791	78	21 158	125.2
2014	10 675	8898	48	19 621	113.6

Source: Adapted from London Fire Brigade Information Management Team

Look at the data source

1 Look at the data for the three decades: the 1980s, the 1990s and the 2000s.

    a Calculate the minimum, lower quartile, median, upper quartile, maximum and interquartile range for the total number of fires per year in each decade.

    b **i** Use your information to draw a box plot for each decade. Draw the box plots one above the other on the same scale for easy comparison.

      ii Comment on your box plots.

    c Use appropriate calculations to determine whether the data contains outliers.

Q1 hint

Which years should you ignore?

2 **a** Calculate the mean and the standard deviation.

b Comment on the mean and the standard deviation.

3 The product moment correlation coefficient (PMCC) between the number of primary fires and the number of secondary fires is 0.805.

a Describe the level of correlation between the number of primary fires and the number of secondary fires.

b Explain why there is limited causation between these two variables.

4 **a** Plot a scatter diagram of the number of chimney fires against year.

b Does it look as if the data would fit a linear model? Explain your answer.

Nicky and Meshaal suggest different models for the data, using the variables

x for the year, with 1980 being year 1, 1981 being year 2 and so on

y for the number of chimney fires.

Nicky suggests using a power model to represent the data. She works out the formula

$$y = 1621x^{-0.924}$$

Meshaal suggests using a quadratic model to represent the data. He works out the formula

$$y = 0.7205x^2 - 41.647x + 648.68$$

c Use Nicky's and Meshaal's models to predict the number of fires in the following years.

Year	Power model	Quadratic model
1980		
1985		
1990		
1995		
2000		
2005		
2010		

d Whose model is better? Why?

5 Evan wants to investigate the relationship between the number of fires and the injury rate per million population.

a Explain why it would be better to use the number of primary fires than the total number of fires.

b Calculate the PMCC between the number of primary fires (x) and the injury rate (y).

Use the following values:

$\Sigma x = 652\,372$, $\Sigma y = 6621.2$, $\Sigma x^2 = 12\,617\,890\,920$,
$\Sigma y^2 = 1\,297\,337$, $\Sigma xy = 126\,629\,282.3$ and $n = 35$.

c Describe the correlation between the number of primary fires and the injury rate per million population.

d Do you think it is fair to say there is a causal relationship between the two variables?

e Calculate the equation of the least squares regression line.

f Interpret the least squares regression line.

g The PMCC between the number of secondary fires and injuries per million people is 0.533. Explain why this is a meaningless statistic.

Explore

The number of primary fires was relatively constant from 1980 until 2003, and then they started to fall sharply. Can you find any evidence of fire prevention advertisements around this time?

Secondary fires are often outdoor fires and most commonly occur in the summer, when the weather is hotter and the ground is drier. Can you find any link between weather and the number of secondary fires?

There are some noticeable spikes in the number of secondary fires, for example in 1995 and 2003. Were these years warmer than average?

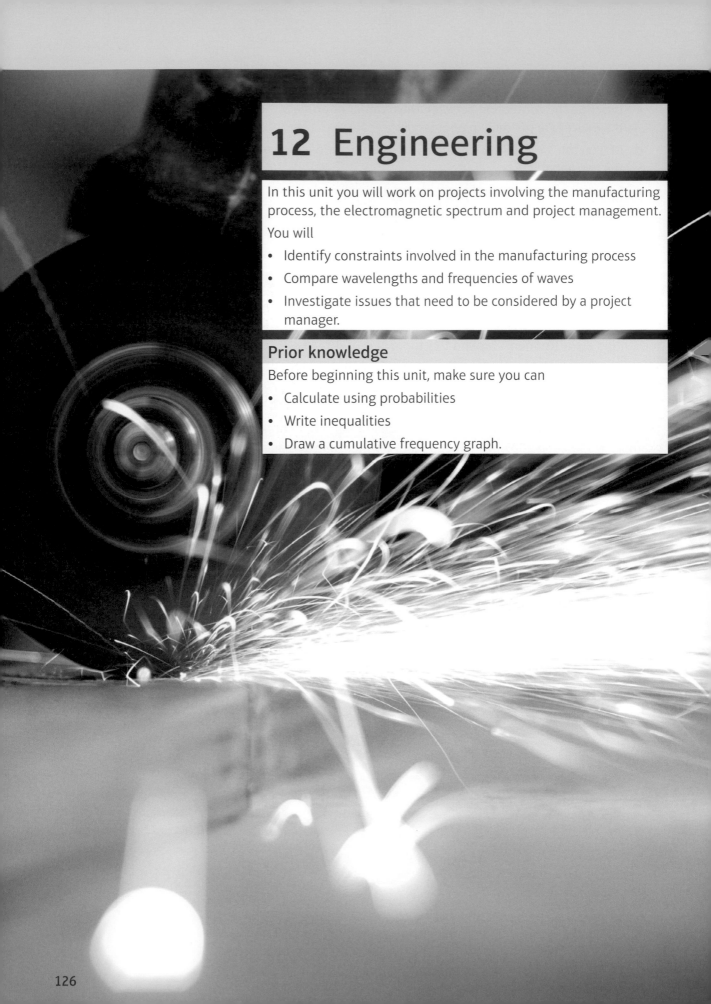

12 Engineering

In this unit you will work on projects involving the manufacturing process, the electromagnetic spectrum and project management.

You will

- Identify constraints involved in the manufacturing process
- Compare wavelengths and frequencies of waves
- Investigate issues that need to be considered by a project manager.

Prior knowledge

Before beginning this unit, make sure you can

- Calculate using probabilities
- Write inequalities
- Draw a cumulative frequency graph.

12.1 Making paper

Data source A

A manufacturer produces paper for the newspaper industry. It has a contract with a major newspaper to produce newsprint and glossy paper for the magazine insert in the Sunday edition.

The machine used to manufacture the paper runs at 1650 metres per minute for newsprint and 2000 metres per minute for magazine-quality paper. Both types of paper are 10 metres and have a mass per unit area of 45 grams per square metre.

In order to stabilise the material while it is going through the process, a special polymer must be added. Newsprint requires 500 g of polymer per tonne of material and magazine-quality paper requires 800 g per tonne. The total amount of polymer available per week for combining with the material is 4 tonnes.

The manufacturer wants to maximise its profit. The profit on newsprint is £100 per tonne and on magazine-quality paper is £195 per tonne.

As the demand for magazine-quality paper is less than for newsprint the manufacturer ensures that no more than 20% of the total weight of material produced should be magazine-quality.

The machine runs for 150 hours per week.

Data source B

Operations managers are expected to monitor and improve the efficiency of the manufacturing process by ensuring that Overall Equipment Effectiveness (OEE) is as high as possible. OEE is based on three metrics: availability, performance and quality.

- Availability is the percentage of the total production time available that the machine is running.

- Performance is the actual amount of material produced, as a percentage of the total amount the machine could produce.

- Quality is the amount of acceptable material produced, as a percentage of the actual amount produced.

Table 1 gives the generally accepted goals for each of these metrics, together with the overall resulting OEE. There are many references to the world class OEE being 85%. It is a convenient, compelling and completely artificial benchmark. Companies should set their own OEE targets that will drive solid, incremental improvement.

Technical literacy

A metric is a measure used to assess the performance of a business.

Table 1 World class overall equipment effectiveness (OEE) metrics

OEE metric	World class value
Availability	90.0%
Performance	95.0%
Quality	99.9%
OEE	85.0%

Source: www.oee.com

Table 2 is a summary of the paper manufacturer's performance on each of the three metrics for the 50 weeks' operating period during one year. (There is an annual two-week shut-down for essential maintenance.)

Table 2 Manufacturer's performance against OEE metrics

Value of metric	Number of weeks this percentage was achieved		
	Availability	Performance	Quality
$88\% \leqslant x < 90\%$	1	1	0
$90\% \leqslant x < 92\%$	4	2	0
$92\% \leqslant x < 94\%$	11	3	1
$94\% \leqslant x < 96\%$	18	18	3
$96\% \leqslant x < 98\%$	12	19	18
$98\% \leqslant x < 100\%$	4	7	28

Look at the data sources

Using Data source A

1 The rate at which newsprint can be produced is 44.55 tonnes per hour.
Calculate the number of tonnes of magazine-quality paper that can be produced in an hour.

Q1 hint
Rate of production = speed × mass per unit area
1 tonne = 1000 kilograms

2 Let x be the number of tonnes of newsprint produced in one week and y the number of tonnes of magazine-quality paper produced in one week.
Write a formula for the objective function, P.

Q2 hint
The objective function defines whatever it is that must be optimised.

3 The inequality which represents the constraint on the total machine time is $\frac{x}{44.55} + \frac{y}{54} \leqslant 150$
Write the inequalities to represent each of these constraints:

a The total amount of polymer available per week

b The maximum amount of magazine-quality paper.

4 **a** Represent the constraints on a graph.
Shade the area which does not satisfy the inequalities.

b Draw lines on your graph to show that the number of tonnes of newsprint and magazine-quality paper cannot be negative.

5 Which values of x and y will give the maximum income?

6 The company changes the 20% constraint on the magazine-quality paper. How would this affect the possible profit?

Using Data source B

7 Construct a compound histogram for availability, performance and quality.

8 Does the manufacturer comply with the suggested OEE standards? Explain your answer.

9 Draw a suitable graph for each metric to work out the median and upper and lower quartiles.
Does this graph support your findings from Question 8?

Q7 hint
In the same way that you have constructed compound bar charts, you can construct compound histograms.

Construct a histogram for the availability data first, then construct the bars for the performance and quality on top of the availability bars. Check the total height of the compopund bar is equal to the total frequency density for all three metrics.

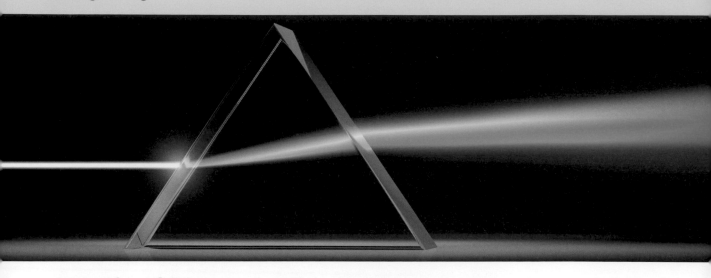

12.2 The electromagnetic spectrum

Data source A

Electromagnetic waves can be described in terms of three variables:

Wavelength

The wavelength is the length of one complete wave. It is the distance from any point on the wave to the place where the pattern repeats.

Frequency

The frequency of a wave is the number of waves produced by a source each second. It is also the number of waves that pass a certain point each second. Frequency is measured in hertz (Hz). For very high frequencies, you can use kilohertz (kHz), megahertz (MHz) and gigahertz (GHz).

Amplitude

The amplitude of a wave is the maximum displacement (the height of a crest or the depth of a trough) that a point moves away from the rest position.

Figure 1 illustrates the amplitude and the wavelength.

Figure 1 An electromagnetic wave

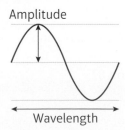

The velocity of transmission of a wave is equal to the wavelength multiplied by the frequency:

$$v = f\lambda$$

where v is velocity

f is frequency

λ (lambda) is wavelength.

Data source B

As physicists investigated the new discoveries in electricity in the 19th century they realised that the interaction of electric and magnetic fields would produce waves. The Scottish physicist James Clerk Maxwell first worked out the equations that described these waves, and then the German physicist Heinrich Hertz first produced electromagnetic waves that conformed to Maxwell's equations.

The electromagnetic spectrum covers a wide range of types of waves, of which visible light is one part. Table 1 lists these types of waves, from very short wavelengths of about one trillionth of a metre to very long wavelengths of over a kilometre. Table 2 lists the internationally accepted (*Système International*, SI) prefixes for units as used in Table 1.

Table 1 The frequency and wavelength of different types of electromagnetic waves

Radiation	Typical frequency	Typical wavelength
Gamma rays	300 EHz	1 picometre
Hard X-rays	10 EHz	30 picometres
Soft X-rays	3 EHz	100 picometres
Ultraviolet	30 PHz	10 nanometres
Visible – violet	710 THz	400 nanometres
Visible – red	450 THz	720 nanometres
Infrared	30 THz	10 micrometres
Microwaves	30 GHz	1 centimetre
Radio – UHF	3 GHz	10 centimetres
Radio – VHF	30 MHz	10 metres
Radio – AM shortwave	10 MHz	30 metres
Radio – AM medium wave	1 MHz	300 metres
Radio – AM long wave	200 kHz	1500 metres

Technical literacy

UHF, VHF and AM are all different types of radio waves.

Table 2 SI prefixes for hertz (Hz)

Submultiples			Multiples		
Value	SI symbol	Name	Value	SI symbol	Name
10^{-1} Hz	dHz	decihertz	10^{1} Hz	daHz	decahertz
10^{-2} Hz	cHz	centihertz	10^{2} Hz	hHz	hectohertz
10^{-3} Hz	mHz	millihertz	10^{3} Hz	kHz	kilohertz
10^{-6} Hz	µHz	microhertz	10^{6} Hz	MHz	megahertz
10^{-9} Hz	nHz	nanohertz	10^{9} Hz	GHz	gigahertz
10^{-12} Hz	pHz	picohertz	10^{12} Hz	THz	terahertz
10^{-15} Hz	fHz	femtohertz	10^{15} Hz	PHz	petahertz
10^{-18} Hz	aHz	attohertz	10^{18} Hz	EHz	exahertz
10^{-21} Hz	zHz	zeptohertz	10^{21} Hz	ZHz	zettahertz
10^{-24} Hz	yHz	yoctohertz	10^{24} Hz	YHz	yottahertz

Most of these types of electromagnetic waves are invisible to the human eye. Electromagnetic waves are only visible light between ultraviolet (UV) and infrared waves. Within this visible light are all the colours of the visible light spectrum.

Table 3 lists the frequencies and corresponding wavelengths of different colours within the visible spectrum. Table 4 shows the values for the same variables for selected radio stations, in order of decreasing wavelength.

Table 3 Frequencies and wavelengths within the visible spectrum

Colour	Frequency (THz)	Wavelength (nm)
Red	455	660
Orange	500	600
Yellow	526	570
Green	556	540
Blue	625	480
Violet	714	420

Table 4 Frequencies and wavelengths for radio stations

FM station	Frequency (MHz)	Wavelength (m)
Radio 2	90.0	3.33
Radio 3	92.4	3.25
Radio 4	94.4	3.18
CSR FM	97.4	3.08
Radio 1	98.8	3.04
Classic FM	101.8	2.95
Heart	102.7	2.92
Radio Kent	104.2	2.88
KMFM	106.0	2.83

Look at the data sources

Using Data source A

1 **a** Choose one of the types of electromagnetic waves.
Work out its velocity, to 1 significant figure.

  b What is the velocity of the other types of electromagnetic waves?

2 Show that the relationship between frequency and wavelength is a reciprocal relationship.

Using Data source B

3 Plot a graph of wavelength against frequency for light within the visible spectrum.

Q2 hint

For a reciprocal relationship, $y = \dfrac{k}{x}$, where k is a constant. This can be written as $xy = k$.

Explore

Are radio waves light waves? Explain.

12.3 Project management

A construction company is preparing to bid on a contract to build a new sports hall for the local college. A project manager for the company considers the stages of work that will be required to build the hall.

The first stage involves clearing the site and carrying out the groundworks. The manager estimates that this will take between 2 and 4 months. The second stage is the construction of the building and the internal fittings, including electric, gas and water supplies. This is estimated to take between 6 and 8 months.

The manager analyses similar projects the company has completed in the past and estimates the probabilities for completing the stages are as follows:

Stage 1   Probability of stage taking 2 months = 0.5
           Probability of stage taking 3 months = 0.4
           Probability of stage taking 4 months = 0.1

Stage 2   Probability of stage taking 6 months = 0.3
           Probability of stage taking 7 months = 0.5
           Probability of stage taking 8 months = 0.2

The college wants Stages 1 and 2 completed within 10 months. It is willing to pay a bonus if the work is completed early and will charge a penalty if it is not finished within 10 months.

The bonus is 5% of the fee for each month that the project is completed early. The penalty is 15% of the fee for each month that it overruns.

The company decides to submit a proposal of £800 000 to carry out the work for Stages 1 and 2. This quote estimates a profit of £100 000.

Look at the data source

1   Prepare a tree diagram to enable you to calculate the probability that the job will be completed within 10 months.

2   List the probabilities of all the possible outcomes of the job, from the shortest time it could take to the longest time.

Q2 hint

Check that the probabilities add up to 1.

3   What is the key assumption that you made in order to calculate the probabilities in the tree diagram?

4   Estimate the cost to the college for each scenario.

5   Estimate the profit (or loss) for the construction company for each scenario.

6   Calculate the expectation value of the cost to the college.

Q6 hint

The expectation value of a quantity is the sum of each of the possible values it can attain multiplied by the probability of that value occurring.

7   Calculate the expectation value of the profit (or loss) to the construction company.

8   The company's policy is that it will not bid on a contract if the expected profit is less than 10% of the contract cost. Should it go ahead with bidding for this job?

Index